I0469420

PERFECT ASTROLOGY (SYNASTRY)

Author: Shri Ram Babu Sao

Shri Ram Babu Sao is meritorious, genius and very intellectual person since childhood and is a Mechanical Engineer (1970) from Patna University. He secured the "National Merit Scholarship" of India during Secondary Board Examination. He has deep interest in Astrology since childhood. He is deeply committed and involved in Astrological research and development work. Sincerity, deep study and his hobby has put him into the use of many new techniques and methods for horoscope prediction. He has studied many books and magazines on astrology. He thought of the necessity of a consolidated Book covering all the aspects, topics on subject matters pertaining to astrology at one place. This Book "Perfect Astrology" is the jest of most of the topics on astrology. This will result in improving the general quality levels of the reader to a greater satisfaction. He has taken a lead in upgrading the process of awareness of various matters benefiting the fresh learner in astrology. Any body can avail of practical knowledge on various topics related to astrology and can make predictions in detail of himself or his family member with the help of this Book.

Contact: Mobile: 9819506068
                Email: rbsao8844@yahoo.co.in ;

Disclaimer

The book "PERFECT ASTROLOGY (SYNASTRY)" is not a writer's whole & sole product. It is a combination of the knowledge and expertise of the author and the data collected from different Books, specially researched to meet the objective and to enhance the knowledge on Astrology. Wherever necessary, the reference of the above books has been given. Because the prediction is not based on the contents of the book but on the skills, purity and perfection of the astrologer, any prediction based on this book shall be the responsibility of the person making Synastry.

ISBN-13: 978-1530059553

ISBN-10: 1530059550

First Edition: 2016

Publisher:
Amazon.in

Preface

This Book "PERFECT ASTROLOGY (SYNASTRY), is a unique book, which is very informative and also easy to understand. One book is truly equivalent of several books on astrology of Synastry for making the best and long lasting relationship or Partner. You must make synastry about of your partner before getting involved into business or life. Synastry is the art of relationship Astrology. It is a fascinating and illuminating study of how individuals interact with one another. Each individual is born with a personal birth chart, which is a map of the heavens. The birth chart has the effect of stamping, or imprinting, the energies of the planets and signs on an individual. When we interact with others, the individual energies of our natal charts form special relationships with the individual energies. Many of us are familiar with the study of Sun Sign Compatibility. Some will ask, for example, "Does a Leo get along with a Scorpio?" While these comparisons have some value, they are very general. Many other factors are involved when evaluating the compatibility of two people. Although Synastry is complex, have tried to turn to some especially useful methods that how you will have relationships with your business or life partner. This book will shed some light and will compare two or more Natal charts in order to analyze and forecast the interaction of the individuals involved. "Money is Prosperity and before making money you must make synastry about your partner with the help of this book, which provides you great tool in the form of a Nakshatra, Moon Sign, Ascendant, Ashtakuta and the "Sun Sign" Synastry of your business or life partner with the help of this book. This provides some of the elementary and in depth essential elements on complete knowledge. Many of the basics on astrology are explained in detail.

The sun, which is actually a star, is at the centre of the Solar System, and likewise, the Sun Sign describes our unique personal centre. It's the core of the individual; the overriding nature and will of the individual. This describes "us" and is our "awake" side, our consciousness and outer-directed individuality. Thus, the Sun is the most important single indicator, fuelling the total personality. The nature of our Sun sign is our true colours that we display to the world day in and day out. Our zodiac sun sign signifies the energy on which our

Soul is incarnated for lifetime. Though there are numerous additional factors influencing the overall map of our Soul, our sun sign plays a major role in our life. Our sun sign reveals our consciousness of self as a unique human being. The self we project to the world is most often a perfectly blend of your sun sign. In soul-cantered astrology, the Sun in its sign has four primary purposes:

- It reveals the energies, temperament, and characteristics, nature, behaviour and overall personality of the individual.
- It speaks about the nature of the personal life and talents in order to be judged by others.
- It is the force field through which the individual's personality expresses itself.
- It reveals certain indications about the makeup of the personality.

The Sun Sign is the core identity, the very essence of the basic vitality of an individual. Sun sign gives valuable information about the general themes for everyone. The business partnership events become clear by Sun sign horoscope. The Sun sign is just the tip of the iceberg in Astrology, the study of which can take us deep into the mysteries of any individual. There are twelve Sun signs, each with its character traits, inclinations, leanings, motivations, etc. The sun sign blends with the rest of our planets to form your unique, one-of-a-kind astrological blueprint. Sun sign describes our basic nature and the personality traits that remain constant through the ups and downs of life. It is the image we shine out to the world. In romance, marriage or Business Partnership, knowing how a certain Sun Signs behave at close range can save you a lot of heartache or help you plan our strategy in life. The sun sign gives crucial clues to understanding the dating behaviour of the individual or opposite sex. Sun Sign is the centre and active part of our personality.

Astrology is planning your future by averting the misshapenning by action in the right time by synastry. The whole secret of Astrology is "Right Timing". By using the book, your life will be more prosperous than ever before. It is important to work "Smartly" but not hard. Many people need to know about their financial status, most important events and their future whether it is going to improve or is it going to get worse. Is now the time to spend or save? Money controls the way we live our lives, and this amazing readings will give you the insight, you need to take control of your financial future. It could be one of the most important readings that you will get through this Book that can truly change your life. This Book will isolate time to capture the situation and reveal its significance. At last, you have discovered a direct channel that will allow you an insight into your own destiny. Many "psychic services" charge you as much as $50, $75, or even $100 for a single reading, but, this book offers you a single instrument for reading as many as you want and that too at no cost. The technically advanced matters allow you to deliver your reading to you quickly and effectively. Not only will your reading be incredibly accurate but also you will have it available to read and analyse at your own pace. In addition, this book offers you an opportunity to record your own readings and readings of your family members by yourself. Just realise that how much you are going to save on account of Horoscope readings for you and your family members.

This is your journey. You may feel compelled to consult the any Online Psychic frequently, or as often as your circumstances warrant. Through this book, you may record the information you read and so you can review, re-evaluate and reconsider its significance as often as you like.

As you continue to consult this book for answers to important issues in your life, an accurate record of your responses will be chronologically archived to allow you to a greater understanding of yourself as you relate to the complex world around you. Gain Wisdom through knowledge of planetary influences by understanding with whom and when to act or not to act. Gain Power and Status using your own cycles that empower you toward success by knowing your marriage partners, Business Partners, or Best Friends. Avoid Problems before they happen by knowing the inbuilt nature of your marriage partners, Business Partners, or Best Friends before making them partners of friends. Be Comfortable with gifts provided by the universe, as you deserve them.

Enormous books are available in the market. One has to read many books to master on Sun Sign predictions and also to predict the individual nature just with the help of his birth date. Our life is speedy. It is ever active and is changing every moment. Each one of us is facing difficulty or being cheated by the partners in business at every step. This book will facilitate to reach your destination by moving ahead with ease even in the storming situation. This is so much strife and struggle in the present time and most of the people are cheater as it was never before. This is a time of ready-made food and fast food. Nobody has time to cook the food and then eat. Only this feeling motivated me and necessitated writing this book. This is easily approachable and compact. It is full of all information at one place to be referred easily and quickly by anybody whether busy in any profession, particularly in business or marriage alliance.

Today the eyes of the whole of the world are fixed on India for any kind of development, because our India Astrology is the best in the world and covers almost every aspect of the individual life along with the timings of the events to happen in individual life. Now it is a duty of an Indian to come forward to lead and direct the world in this sphere and let the world know how resourceful Indian are in astrology. I am confident that this book will help to all in achieving his object and success in every field of astrology. I have tried to make clear about what is the correct astrology work. These are all correct and true facts & figures collected from various books and incorporated here in a single book for the first time for use by the common men. Behind all this, there is our exhaustive study and collections. More than the study is the presentation of the subject matter and even much more than the presentation of the subject matter is long years of experience and association with the astrology work all over India and abroad. This gives the authenticity to the book. This book is an asset for the Jyotish and the Professionals too.

The recognition and importance, which this book has received within a short time, is a positive proof of its efforts and impartiality. Hundreds and thousands of persons from our country and from foreign countries are responding and referring this book. There is a great demand created for this book. This is no credit to me but is the genuineness and absolutely authenticity and clarity of this book. I am confident that the readers and experts will consider my effort.

Content

1

Introduction

1.1 History of Sun Sign

The growth and achievement in life of a Mankind (Individual) depends on so many factors, such as, 1) Ascendant, Sun, Moon and Stars (Nakshatra) astrological Effects, 2) Genetic Effects, 3) Environmental Effects, and last but not the least, 4) Society & Contacts. Astrological Effects can be found out by "Astrology" or "Jyotish" which means the 'science of light' and is related with the Light and Magnetic Field emitted by the planets (Graha). Indian Astrology has been divided into three main branches of study.

1. Siddanta,   2. Samhita,   3. Hora

Siddanta: Siddanta covers astronomical study of celestial bodies.

Samhita: Samhita deals with mundane astrology such as earth quakes, floods, volcanic eruptions, rainfall, weather conditions economic conditions and effects of sunspots.

Hora: Hora Astrology deals with the Phalitha Jyotish, which means predictions about individual's life. Hora Astrology has six sub-divisions, namely, Jathaka, Gola, Prasana, Nimitta, Muhurata and Ganitha.

'Hindu Astrology' is founded by the Maharashi Aryabhatta, Parasara, Varaha Mihira, Jaimini, Garga, Kalidasa and Kalyan Varma. The origin of this science can go back as old as 4000 years. The astrology fully knows the individual's future as indicated by horoscope but can't certify the same as it is not 100 percent Mathematics or Science".

The importance of Sun Sign, Lagna sign, Moon Sign and Stars (Nakshatra): In Vedic Astrology, Sun Sign, Moon Sign, Lagna sign and Stars (Nakshatra), are the tools to find a perfect partner who suits him. In astrology, it is assumed that certain people sign or Stars fits to people of the other's specific zodiac sign. The zodiac is assigned with certain characteristics, energies and spirits of this universe. They give life or essence to our reality and are the forces behind all material manifestation. Each of the Sun Sign, Lagna sign, Moon Sign and Stars (Nakshatra) contains or encapsulates a group of principles for the functioning of life and in turn influences the physical and spiritual centres (chakras and glands) of mankind on the Earth. The chakras in the human body, such as, Moon chakra or Sun chakra signifies different stages of our development, from past to present to future. Zodiac Signs have four the most basic elements (fire, earth, air, and water) and their corresponding principles of spirit, matter, mind, and soul match or mismatch with each other. Each sign has one of four associated elements attached to it:

Fire Signs: Aries, Leo and Sagittarius.
Air Signs: Gemini, Libra and Aquarius.
Earth Signs: Taurus, Virgo and Capricorn.
Water Signs: Cancer, Scorpio and Pisces.
Your best Partner is always as mentioned below:
1. Same Sign
2. Same Element
3. Compatible Elements
In general, you should be compatible with people of your own zodiac sign, e.g. Aries with a fellow Aries, as well as any sign that falls within your particular element. Example: If you are a Virgo (Earth element), then your best matches would be with other people who is Virgo or which fall under the Earth element like Taurus and Capricorn.

While the same sign or same element is generally your best compatible Partner, you may also have great relationships with compatible element signs. Example: Air and Fire signs work well together because Fire needs Air. I'm a Libra (Air) and my wife is a Leo (Fire) and we have an amazing relationship with one another. These types of relationships are generally strong but some may be too polar opposite to be compatible with one another.

Sun Sign: Sun Sign is the active part of your personality and shows itself with blaring intensity, behaviours, and the personality traits. Moon Signs convey our shadow selves; personality traits, but shown through our Sun Signs are bold and clear.

Sun Sign	From	To	Element
Aries	March 21st	April 19th	Fire
Taurus	April 20th	May 20th	Earth
Gemini	May 21st	June 20th	Air

Cancer	June 21st	July 22nd	Water
Leo	July 23rd	August 22nd	Fire
Virgo	August 23rd	September 22nd	Earth
Libra	September 23rd	October 22nd	Air
Scorpio	October 23rd	November 21st	Water
Sagittarius	November 22nd	December 21st	Fire
Capricorn	December 22nd	January 19th	Earth
Aquarius	January 20th	February 18th	Air
Pisces	February 19th	March 20th	Water

Moon Signs: Moon Signs help define our emotional development and express the unconscious side of our personality. Moon is said to represent your instinctual self, which many people keep hidden. The Moon also influences our senses depending on its placement in the birth chart. Most people will have their Moon in a Sign that is different from their Sun. This is why many people with the same Sun Sign can be so different from the same definition given by Sun Sign. Moon Signs may be a much more accurate description of what a person is like.

Nakshatra: Nakshatra is the term for lunar mansion in Hindu astrology. A Nakshatra is one of 27 sectors along the ecliptic. Their names are related to the most prominent asterisms in the respective sectors.

Ascendant: The Ascendant is said to be the mask one wears in public and describes the aspect of personality traits you exhibit, influences physical characteristics, image, style and mannerisms and how one act.

In classical Jyotish the Moon has equal or greater power to the lagna (Ascendant). Accurate readings cannot occur from the Rashi-Lagna only. Accurate readings require at least two preliminary scans (predictions): first from the Rashi-Lagna, and second from Chandra-Lagna. Then more detailed scans such as from the Mahadasha-Lord and the Karaka are also required. More accurate Jyotish predictions are often produced by reading significations from the Chandra Lagna first, Radical Lagna second, Navamsha Lagna third, and Varga Lagna additionally. Thus it is easier to identify the behaviours and environments that provide the best psycho-emotional support while predictions of a Horoscope by signs. Horoscope is a mirror in which an astrologer can see one's past, present and future. Horoscope is like a snapshot of a particular place in time and space. For casting the natal horoscope of an individual the time of birth, date of birth and place of birth is needed. There are 12 houses and 12 Signs in a horoscope from which an astrologer can predict about various areas of the life of an individual. This "PERFECT ASTROLOGY (SUN SIGN)" Book enables the astrologer to know that what the future has in store for the native.

The sun, which is actually a star, is at the centre of the Solar System, and likewise, the Sun Sign describes our unique personal centre. It's the core of the individual; the overriding nature and will of the individual. This describes "us" and is our "awake" side, our consciousness and outer-directed individuality. Thus, the Sun is the most important single indicator, fuelling the total personality. The nature of our Sun sign is our true colours that we display to the world day in and day out. Our zodiac sun sign signifies the energy on which our Soul is incarnated for lifetime. Though there are numerous additional factors influencing the overall map of our Soul, our sun sign plays a major role in our life. Our sun sign reveals our consciousness of self as a unique human being. The self we project to the world is most often a perfectly blend of your sun sign.

In soul-cantered astrology, the Sun in its sign has four primary purposes:

- It reveals the energies, temperament, and characteristics, nature, behaviour and overall personality of the individual.
- It speaks about the nature of the personal life and talents in order to be judged by others.
- It is the force field through which the individual's personality expresses itself.
- It reveals certain indications about the makeup of the personality.

The Sun Sign is the core identity, the very essence of the basic vitality of an individual. Sun sign gives valuable information about the general themes for everyone. The business partnership events become clear by Sun sign horoscope. The Sun sign is just the tip of the iceberg in Astrology, the study of which can take us deep into the mysteries of any individual. There are twelve Sun signs, each with its character traits, inclinations, leanings, motivations, etc. The sun sign blends with the rest of our planets to form your unique, one-of-a-kind astrological blueprint. Sun sign describes our basic nature and the personality traits that remain constant through the ups and downs of life. It is the image we shine out to the world. In romance, marriage or Business Partnership, knowing how a certain Sun Signs behave at close range can save you a lot of heartache or help you plan our strategy in life. The sun sign gives crucial clues to understanding the dating behaviour of the individual or opposite sex. Sun Sign is the centre and active part of our personality.

1.2 Astrological Terms

Ascendant (Lagna): The Ascendant or Lagna or Rising Sign is the Sign in which an individual is born.

Angle (Kendra): The Ascendant (first house), Descendant (fourth house), Mid-heaven Seventh house) and I Mum Collie (tenth house) are called Kendra (Angle).

Aspect (Drishti): A planet affects another planet or house with its full magnetic power as if he is present there and is called the Aspect (Drishti) of the planet. This is the angular relationship between two planets or house and the planet by their position.

Association: It is a relationship between two or more planets by their position in a house.

Auspicious Planet: Planet, which gives positive effects and good result to the individual is called auspicious planet.

Barren Signs: Gemini, Leo and Virgo are associated with infertility and are considered Barren Signs. Aries, Sagittarius and Aquarius are considered semi-barren Signs.

Benefic Planet: Some of the planets are considered positive for the individual giving good effects and influences and hence they are called Benefic Planet.

Bhava (House): The complete Zodiac is divided into twelve parts for the purpose of complete study of astrology. Each division is called a Bhava (House).

Birth Time: This is the moment of first breath of a new born individual.

Chart: It is a figure or sketch consisting of 12 houses, in which the position of the planets and the Signs are given.

Combustion (Astangatha/Vikala): The planet associated with an identical longitude or equal longitude or in Conjunction to the Sun within a certain orb or reaching nearer to the Sun by the distance mentioned in the degree, such as, Moon-12^0, Mars-17^0, Mercury-13^0, Jupiter-11^0, Venus-9^0, and Saturn-15

^{0} is called Combust is called Combustion. The energy of that planet is considered burnt by Sun and the planet does not operate independently any more. It is true that a combust planet is week.

Conjunction: Two planets situated together in a house or occupying position close to each other within a certain orb or reaching nearer are called in Conjunction.

Debilitation (Khala): When a planet occupies its Sign of Fall for Neecha Bhanga, the planet is in Debilitation and is considered weak or Neecha.

Degrees of Maximum Exaltation & Debilitation: Maximum degree of Exaltation or Debilitation is the defined degree of planet position in a Sign.

Dignity: A planet is dignified when it occupies its Own Sign, its Moolatrikona or its Exaltation Sign, aspect by a benefic planet without any aspect or affliction by a malefic, when it is not retrograde or when it is increasing in its light. Such planet gives good results.

Direct Motion: It is a motion of the planet that follows the natural order of revolution cycle. The letter "D" marked against a planet indicates the direct motion of planet.

Element: Twelve Signs have been divided into four groups of each three Signs having same nature, called Elements. These Elements are called Fire, Earth, Air or Water.

Exaltation: The planet in a particular Sign is called the planet in Exaltation Sign. Planet is dignified during his Exaltation.

Friendly Planet: The planets of one group are called friendly planets.

Functional benefic Planets: As per Lagna Sign of the Natal Chart, some planets are defined as Functional benefic Planets even though they are Natural Malefic. Their influence is thought to be positive or constructive for the native. It makes the house strong in which he is sitting. The planet gives good result during his Main Dasa and the Antar Dasa fouling together in the same period.

Functional Malefic Planets: As per Lagna Sign of the Natal Chart, some planets are defined as Functional Malefic Planets whose influence is thought to be negative or destructive for the native. The lord of the 6th, 8th, and 12th are called Functional Malefic (Inauspicious) Planets and will make the house weak in which he is sitting. The planet does not give the good result during his Main Dasa and the Antar Dasa fouling together in the same period.

Group of Planets: There are nine planets. They are divided in two Groups, such as, Jeeva Group: The Sun, Moon, Mars, Jupiter and Ketu and the signs that they rule, are in Jeeva Group with the Jupiter as its leader. Sareera Group: The Mercury, Venus, Saturn and Rahu and the signs that they rule, are in Sareera Group with the Saturn as its leader. Both groups are equally important. The planets and their Signs (Rasi) of one group are favourable to each other planets and Signs in that kundali. At the same time, the planets and Signs of one group are inimical (having enemy's behaviour) to the other group's Planets and Signs. In special case, when the planets of one group are harmoniously related to the planets of other group due to other governing factors of the astrology in the Kundali, then there will be the Rajya Yoga in the life of that native.

Gochara (Transit): The revolution or usual movement of planets in the Zodiac Sign during the course of revolution around the Sun is known as Gochara (Transit).

Graha Yudha (planet war): Whenever one planet comes within 5^0 orb of another one among the planets Mars, Mercury, Jupiter, Venus and Saturn, as viewed from the Earth, it causes planetary war, or Graha Yudha. One of the two planets involved in this war is said to be vanquished and another is a victor. The victorious planet produces powerful auspicious effects, while the vanquished or defeated one becomes inauspicious. The house in which this phenomenon occurs is destroyed and the individual suffers throughout his life with respect to that house events.

Horoscope: It is the Janma Kundali, which depicts the positions of different signs and planets at the time of birth. It also represents the Rising Sign at the place of birth and the location of planets in various signs.

Janma Nakshatra: The Nakshatra ruled by the Moon at the birth time of a native is called Janma Nakshatra.

Janma Rashi: Janma Rashi is the Sign occupied by Moon at the time of birth of the native.

Ketu (Dragons Tail/South Node of Moon): Ketu is the dead body of the lusty demon, killed by Vishnu. So, Ketu is the tail of the dragon and called the South Node of moon, which losses and symbolizes the death along with Saturn.

Moon sign: The Moon sign is the zone of the zodiac in which the Moon is positioned when a person is born. This is also called Rashi too.

Natal Chart: The horoscope cast at the birth time of the individual, showing the position of Signs and Planet with respect to house, is called a Natal Chart or Nativity.

Native: It refers to a person (male or female) for whom a horoscope is cast and studied.

Natural Benefic: Moon (waxing), Mercury, Venus, Ketu and Jupiter are called Natural Benefic Planets in order of increasing Benefic.

Natural Malefic: Sun, Mars, Saturn and Rahu are called Natural Malefic Planets in order of increasing malefic. Moon remains a malefic from 9th day of Krishna Paksha to 7th lunar day of Shukla Paksha.

Own house: A planet rules a certain house and that house is called his Own House.

Planet: Sun, Moon, Mars, Mercury, Jupiter, Venus, Saturn, Rahu, Ketu, Uranus, Neptune and Pluto are the twelve heavenly bodies which appear to move in the Zodiac and influence the human body and are called Planets.

Prediction: Knowing about natives past, present and future with the help of horoscope is prediction. But, in one accident many lives are taken, does that indicate similarity in everyone's horoscope? No, because the horoscope of the place or a vehicle in which passengers are travelling, supersedes the horoscopes of the natives. Hence, in a calamity everyone's horoscope does not necessarily indicate death. However, this

Rising Sign (Lagna): The earth is rotating once a day around its axis. One of the 12 zodiacal signs is entering the 1st house every two hours. The Sign entering into 1st house at the birth time is known as Rising Sign or Lagna.

Ruling Planet: The planet which rules the Ascendant or Lagna Sign is called the Ruling Planet.

Significator: The planet ruling a house is called the Significator of that house.

Signs: The zodiac is divided into twelve parts like Aries, Taurus, Gemini, Cancer, Leo, Virgo, Libra, Scorpio, Sagittarius, Capricorn, Aquarius, Pisces and are called Sign.

Solar Chart: The horoscope Chart with the Sun in the Ascendant is called the Solar Chart.

Sun Sign: In astrology, Sun Sign is defined a period of one month with respect to the person Birth Date and is called Sun Sign. Generally, Sun Sign is 2nd house from the position of Sun in Hindu Astrology.

Trikona (Trine): The Houses 1, 5 and 9 are called Trikona or Trine.

Trikas (Badhakasthana): The Houses 6, 8 and 12 are called Trikas or Badhakasthana. These are the Badhaka houses and considered the evil houses of suffering.

Yoga: The certain planetary combinations are defined as Yoga.

Zodiac: It is literally the circle of stars. Zodiac is defined a band of the heaven approximately 14° wide, centred on the Ecliptic, against which the Sun and other planets are seen to move, as seen from the Earth.

1.3 Astrological Symbol

Sign	Sign Picture	Sign Symbol	Lord & Symbol	Aspect Type & Symbol
Aries		♈	Mars ♂	♂ Conjunction
Taurus		♉	Venus ♀	✳ Sextile
Gemini		♊	Mercury ☿	☐ Square

Cancer		♋	Moon ☽	△ Trine
Leo		♌	Sun ☉	⚯ Opposition
Virgo		♍	Mercury ☿	---
Libra		♎	Venus ♀	---
Scorpio		♏	Pluto ♇	---
Sagittarius		♐	Jupiter ♃	---
Aries		♈	Mars ♂	---

| Taurus |  | ♉ | Venus ♀ | --- |
| Gemini |  | ♊ | Mercury ☿ | --- |

1.4 Kundali (Chart)

A) South Indian style Kundali (Chart): In South Indian Style Chart, the position of the signs is always fixed and the position of the Ascendant is always changing. The houses are counted in a clockwise direction. The upper top left but one Rectangular box, being denoted by the digit 1 is always Aries. The next Rectangular box right to it, being denoted by the digit 2, is always Taurus and so on as written in the Chart. The digit 1 through 12 indicates the position of the Signs fixed in clockwise direction.

12 (Pieces) Venus	1 (Aries)	2 (Taurus)	3 (Gemini) Ascendant Mars
11 (Aquarius) Mercury Ketu	**Lagna Chart-1**		4 (Cancer) Moon
10 (Capricorn) Sun			5 (Leo) Rahu
9 (Sagittarius) Jupiter Saturn	8 (Scorpio)	7 (Libra)	6 (Virgo)

FIG 1: South Indian Style Lagna Chart

It is always fixed in the South Indian style of Chart. The Ascendant (Lagna) falls in one of the Sign depending on rising Sign and is called Lagna or Ascendant of the Chart. The counting of the Houses is always done in clockwise direction from the Ascendant (Lagna) as first house through twelfth house. The planets occupy the House according to their longitudinal position.

B) North Indian style Kundali (Chart): In the North Indian style Chart, the Houses are always fixed and the rising Sign falls in the 1st House, which is called Lagna or Ascendant, which is at the top in the centre. The Lagna (Ascendant) as well as the other houses are always fixed and are counted in anti-clockwise direction from the 1st House or the Lagna or the Rising Sign. The planets occupy the House according to their longitudinal position. The numbers shown in this format tell us which sign is in the Lagna and other Houses as shown in the Chart.

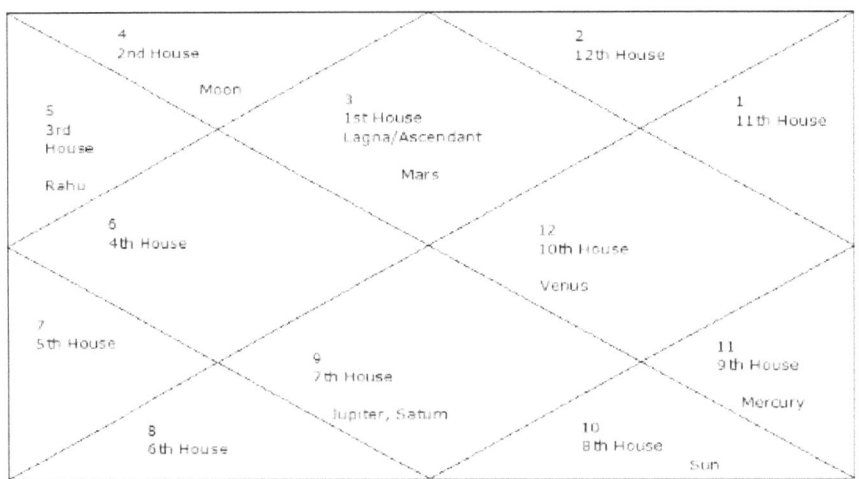

FIG 2: North Indian Style Chart

C) East Indian (Bengali) style Kundali (Chart): In the East Indian (Bengali) style Chart, the Houses are always fixed and the rising Sign falls in the 1st House, which is called Lagna or Ascendant, which is at the top in the centre. The Lagna (Ascendant) as well as the other houses are always fixed and are counted in anti-clockwise direction from the 1st House or the Lagna or the Rising Sign. The planets occupy the House according to their longitudinal position. The numbers shown in this format tell us which sign is in the Lagna and other Houses as shown in the Chart. In Bengali style zodiac is again fixed and ascendant & planets move anti clock wise along the zodiac unlike South Indian System. In western style the ascendant is fixed again and placed on the left hand side whereas

D) East Indian (Bengali) & Western style Kundali (Chart): It is a circular Chart divided in twelve parts in which Sign and Planets position are given as shown below.

Classification Charts: The different type of Horoscope Charts are prepared to study the different aspects of life, such as Natal Chart; Chalit Chart; Transit Chart; Moon Chart; Tithi Parivesha Chart and Divisional Chart

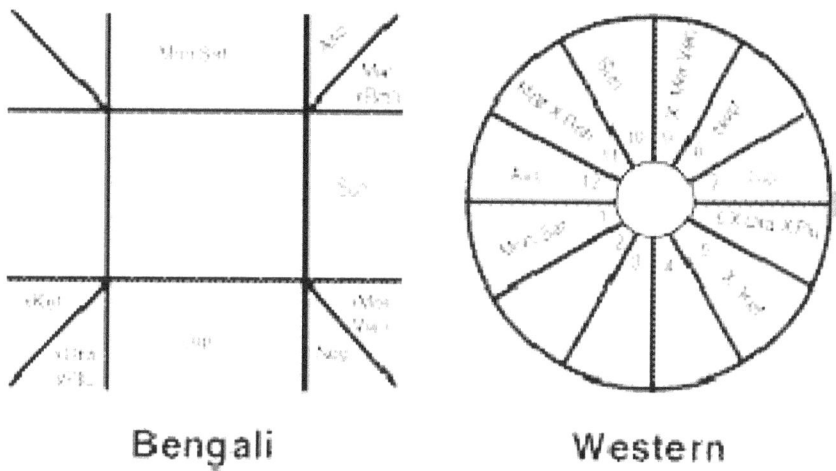

Bengali   Western

FIG 3: East Indian (Bengali) & Western Country Style Chart

Natal Chart: This is the Main Lagna-Chart.
Chalit Chart: Lagna Chart shows the house only approximately with a full Zodiac Sign in it. Chalit Chart is shows the actual Zodiac Sign in a house. Hence Chalit chart is very important to indicate the actual planet position in a particular house and cannot be ignored in astrology. But Chalit Chart shall not be used for determining the aspects or knowing the sign in which a planet is posited or knowing the strength of a planet. For example, if a Sun moves from lagna to 12th house in Chalit Chart while it was in Aries in Lagna chart, then it does not mean that Sun has changed the Sign and shifted to Pisces. It remains in Aries only, but is placed in 12th house because of the actual division of house calculation. Chalit Chart is the actual calculation of the 12 houses of the Lagna Chart and therefore, actual position of planet by making the house division for corrects predictions, because the planets show their behavior according to the house they occupy.

Transit Chart: The Transit Chart or Progression Chart is prepared by positioning the rising Sign in 1st House and planets in the particular fouling Sign at that particular time. Then, this Chat is compared to find the aspect of transiting planet on Natal chart planet to find daily, weakly or monthly forecast of the horoscope when the native wish to find the best time for important event happening such as a change of job, the best time for marriage or having children etc. In short, the Transit Charts are a study aid for planetary influences in the individual life during a particular period.

Moon-Sign (Rashi) Chart: In Moon-Sign (Rashi) Chart, the Sign with Moon placed in the first House as Ascendant and rests other Signs with their Planets follow the Ascendant.

Tithi Parivesha Chart: The Tithi Parivesha Chart manifests concretely the Vimshotari Maha Dasa and Antar Dasa sequence. In this chart, the lord of the Vimshotari Maha Dasa is placed in the first House as Ascendant. In relation to this, the inner workings of the Antar Dasa are revealed with the help of the Antar Dasa lord planet positioned in different Houses. The Antar Dasa Lord brings the timing of the events that come about as a result in life. Tithi Parivesha Chart distils all this along with the transit of the Moon in the Rashi Chakra (Rashi-Chart).

Divisional Charts: In Divisional Charts, each Sign is divided into various divisions, as per the requirements of the type of the Chart. The Divisional Charts are used for study of (1) strength of the planets, (2) important aspects of life. Each Chart gives a clear history of one of aspects of life of the native, such as (1) The physique of the native is known by Lagna Chart, (2) wealth by Hora Chart, (3) happiness through co-born/sibling by Drekana Chart, (4) fortunes from Chaturthamsa Chart, (5) sons and grandsons from Saptamsa Chart, (6) spouse and planet strength from Navamsa Chart, (7) power and position from Dasamsa Chart, (8) parents from Dvadasamsa Chart, (9) pleasure and adversities through conveyances from Shodasamsa, (10) worship from Vimsamsa, (11) learning from Chaturvimshamsa, (12) strength and weakness from Saptavimshamsa, (13) evil effects from Trimsamsa, (14) auspicious and inauspicious effects from Khavedamsa, (15) all indications from Akshavedamsa and (16) Shashtiamsa charts. The Divisional Charts are discussed as follow:

1.5 Reference Books

1. **Hora, Ganita and Samhita:** The great sage Parasara narrated the science of astrology as heard through Lord Brahma in three divisions, viz. Hora, Ganita and Samhita.

2. **Brihat Parasara Hora Shashtra:** The great sage Parasara lived at the time of the Mahabharata war, about 3000 BC. The Brihat Parasara Hora Shashtra (a compendium on astrology) is the primary textbook of Vedic astrology.

3. **Saravali Translated by R. Santhanam:** This focuses on planets in houses and Decanate.

4. **Jaimini Sutras:** The Jaimini sutras by Rishi Jaimini are a unique classic, and considered as next only to the Brihat Parasara Hora Shashtra.

5. **Bhrigu Samhita:** Rishi Bhrigu was the first compiler of predictive Astrology. His famous compilation, Bhrigu Samhita, which contains the predictions for thousands of combinations, is popular even today.

6. **Bhrigu Sutras:** Rishi Bhrigu was the first compiler of predictive Astrology. His famous compilation, Bhrigu Samhita, which contains the predictions for thousands of combinations, is popular even today.

7. **Prasana Tantra:** By Neelakanta Daivagyna around 1550 AD, is a great classic dealing with the Prasana or Horary Astrology and a must for any astrologer.

8. **Brihat Parasara Hora Shashtra Translated by R. Santhanam:** The "bible" of Vedic astrology.

9. **The Inner Sky- by Mr. Forest:** This is a remarkable book. The descriptions of the signs, planets and houses are so effortless that you don't realize that you're learning.

10. **Astrology for Lovers- by Liz Greene:** She is the undisputed queen of psychological astrology. She has captured the essence of each sign of the Zodiac. For example, Aries is brave and Leo is boastful, etc.

11. **Linda Goodman's Love Signs- by Linda Goodman:** She had her finger on the pulse of the signs and a writing style that carries you away to places that few astrologers **can take.**

12. **The Twelve Houses:** This is the best book on the houses, which shows the mission of each house. He talks

about the various signs that rule them and planets being located in them.

13. **Astrology for Yourself-** There is a brilliant step-by-step guide to understanding your chart. You learn about your specific chart and get quizzed at the end of each small, easy to understand chapter.

14. **Astrology for the Soul- by Jan Spiller:** This is a Node book. The information on the nodes has been put here for you in one complete cookbook style reference book.

15. **Chart Interpretation Handbook- by Stephen Arroyo:** He has written a magical little book on interpreting a chart. He respects that you may have your own style and he lays out possible ways that the chart can be interpreted by anyone.

16. **The General Principles of Astrology- by Aleister Crowley:** Long before, he was the most important astrological scholar and authority of the early 20th century.

17. **Saturn a New Look at an Old Devil- by Liz Greene:** She brings profound light to the often-misunderstood Saturn. She explains Saturn by house and by sign giving you a chance to understand the psychological implications it places in your life.

18. **Aspects in Astrology- by Sue Tompkins:** This book is simply fabulous and remarkable gift with wit and wisdom. She manages to illustrate the complexities of aspects with imagery that easily dances in your brain.

19. **Astrology Encyclopaedia- by James R. Lewis:** He has done an impressive job of listing over 780 terms, authors and phenomenon related to astrology.

20. **How to Read Your Astrological Chart- by Donna Cunningham:** He has taken the fear out of interpreting a chart in this delightful, yet insightful cookbook but a dialogue about how to look at the chart. This is a great book for any astrological collection.

2

Asterism (Nakshatra)

2.1 General:

Nakshatra is "constellation" or a group of stars. In Vedic Astrology, Nakshatra is important in predictive astrology and they bring their own properties to the table along with the signs and planets. Nakshatra play an important role in following areas of electional astrology, Vedic Samskara and Samhita:

Nakshatra are extremely important in Muhurta or electional astrology. It is the most important part of the Panchanga that denotes the type of activity that can be taken up when Moon is passing through a specific Nakshatra. There are Nakshatra specified for wedding, tonsuring, starting of education, starting of a business, Namakarana Samskara by specifying the first letter of the name and so on and so forth. Nakshatra forms the basis of horoscope compatibility and matching of charts before marriage or business Partners compatibility. Nakshatra are deemed compatible and incompatible and marriage or partnership is prescribed based on compatibility between prospective husband and wife or two business partners. High level compatibility between Nakshatra is indicated in the detail in the Chapter "Nakshatra" Samhita are based on both Rashi and Nakshatra and are used in mundane astrology predictions. Since the constellation or Nakshatra where moon finds its presence is known as the Janma Nakshatra and as moon influences the mental aspects of the native, Nakshatra of the moon thus casts its indelible influence on moon. The Nakshatra/Moon is connected to our intuition, intelligence and the nature of the mind and emotions. These "Moon parts" within us are ever-changing, fickle, and correlate to the endless movement and change of the Moon. Whereas the movement of the Sun is linked to the 12 signs of the zodiac, the movement of the Moon is connected to the 27 Nakshatra. The Moon is the lord of all the Nakshatra, where the Sun is the lord of all the

Rasi (signs). In the Vedic system of astrology, there are 27 Nakshatra. Nakshatra are Pranic (Sanskrit for life force) in their nature and show the direction of nature's Pranic flow at any given time. The Nakshatra tell us where the Pran is being directed, how its tendencies might manifest, and where Nature is giving her support for expression. They can be seen as natural forces as mental/emotional tendencies or as cosmic archetypes that offer their guidance through elaborate, ancient myths and stories. The most important Nakshatra of the natal chart is one's Moon Nakshatra which represents our mind, intuition and emotional self. The deity will express the quality of a person's thinking and explain their internal processing, while the symbol and planetary ruler ship further refines its potential.

Table 1: Nakshatra Characteristics

Sign	Location of Nakshatra	Nakshatra & its Pad
Aries (Mesh)	0^0 to $13^019'$	Aswini 1,2,3,4 quarters
	$13^020'$ to $26^039'$	Bharani 1,2,3,4 quarters
	$26^040'$ to $30^000'$	Krittika 1st quarter
Taurus (Vrishabha)	0^0 to $10^000'$	Krittika 2,3,4 quarters
	$10^000'$ to $23^020'$	Rohini 1,2,3,4 quarters
	$23^020'$ to $30^000'$	Mrigashira 1, 2 quarters
Gemini (Mithuna)	0^0 to $6^040'$	Mrigashira 3, 4 quarters
	$6^040'$ to $20^000'$	Aridra 1,2,3,4 quarters
	$20^000'$ to $30^000'$	Punarvasu 1,2,3 quarters
Cancer (Karaka)	0^0 to $13^020'$	Punarvasu 4 quarter
	3^020 to $16^040'$	Pushya 1,2,3, 4 quarters
	16^040 to $30^000'$	Ashlesha 1,2,3,4 quarters
Leo (Simha)	0^0 to $13^020'$	Makha 1,2,3,4 quarters
	13^020 to $26^040'$	Purva Phalguni 1,2,3,4 quarters
	26^040 to $30^000'$	Uttara Phalguni 1 quarter
Virgo (Kanya)	0^0 to $10^000'$	Uttara Phalguni 2,3,4 quarters
	$10^000'$ to $23^020'$	Hasta 1,2,3,4 quarters
	$23^020'$ to $30^000'$	Chitra 1,2 quarters
Libra (Tula)	0^0 to $6^040'$	Chitra 3, 4 quarters
	$6^040'$ to $20^000'$	Swati 1,2,3,4 quarters

	20^{0}00' to 30^{0}00'	Visakha 1,2,3 quarters
Scorpio (Vrishchika)	0^0 to 3^{0}20'	Visakha 4th quarter
	3^{0}20 to 16^{0}40'	Anuradha 1,2,3,4 quarters
	16^{0}40 to 30^{0}00'	Jyeshtha 1,2,3,4 quarters
Sagittarius (Dhanu)	0^0 to 13^{0}20'	Moola 1,2,3,4 quarters
	13^{0}20 to 26^{0}40'	Purva Ashadha 1,2,3,4 quarters
	26^{0}40 to 30^{0}00'	Uttara Ashadha 1st quarter
Capricorn (Makar)	0^0 to 10^{0}00'	Uttara Ashadha 2,3,4 quarters
	10^{0}00' to 23^{0}20'	Sharvana 1,2,3,4
	23^{0}20' to 30^{0}00'	Dhanishta 1, 2 quarters
Aquarius (Kumbh)	0^0 to 6^{0}40'	Dhanishta 3,4 quarters
	6^{0}40' to 20^{0}00'	Shathabisha 1,2,3,4
	20^{0}00' to 30^{0}00'	P. Bhadrapada 1,2,3 quarters
Pisces (Meena)	0^0 to 3^{0}20'	P. Bhadrapada 4 quarter
	3^{0}20 to 16^{0}40'	U. Bhadrapada 1,2,3,4
	16^{0}40 to 30^{0}00'	Revati 1,2,3,4 quarters

Table 2: Nakshatra Characteristics

S. No.	Nakshatra Name	Ruler / Lord	Deity / Demigod
1	Aswini	Ketu	Aswini Kumar
2	Bharani	Venus	Yama (Death)
3	Krittika	Sun	Agni (Fire)
4	Rohini	Moon	Brahma
5	Mrigashira	Mars	Moon
6	Aridra	Rahu	Rudra
7	Punarvasu	Jupiter	Aditi
8	Pushya	Saturn	Brihaspati
9	Ashlesha	Mercury	Sarpas (Snake)
10	Magha / Makha	Ketu	The Pitris
11	Purva Phalguni	Venus	Bhaga
12	Uttara Phalguni	Sun	Aryaman
13	Hasta	Moon	Savitri

14	Chitra	Mars	Tvashtri
15	Swati	Rahu	Vayu
16	Visakha	Jupiter	Indragni
17	Anuradha	Saturn	Mitra
18	Jyeshtha	Mercury	Indra
19	Moola	Ketu	Nirriti
20	Purva Ashadha	Venus	Apah
21	Uttara Ashadha	Sun	Vishva
22	Sharvana	Moon	Vishnu
23	Dhanishta	Mars	Vasus
24	Shathabisha	Rahu	Varuna
25	Purva Bhadrapada	Jupiter	Ajaikapada
26	Uttara Bhadrapada	Saturn	Ahirbudhnya
27	Revati	Mercury	Pushan

2.2 Special Nakshatra:

In addition to the above Group Classification of Tara, we have a few special Tara (Nakshatra) for each person: (1) The constellation occupied by Natal Moon is 1st Nakshatra and called Janma Nakshatra. Janma means "birth" and this Nakshatra shows general well-being. (2) The 10th constellation from Janma Nakshatra is called Karma Nakshatra. Karma means "profession" and this Nakshatra shows profession and workplace. (3) The 18th constellation from Janma Nakshatra is called Saamudaayika Nakshatra. Saamudaayika roughly means "related to a crowd" and this Nakshatra shows group activities and friendship. (4) The 16th constellation from Janma Nakshatra is called Sanghaatika Nakshatra. Sanghaatika roughly means "belonging to group" and this Nakshatra shows group/social activities. (5) The 4th constellation from Janma Nakshatra is called Jaati Nakshatra. Jaati roughly means "community" and this Nakshatra shows one's community. To be more correct, one's Jaati shows people who belong to the same class, nature and profession. (6) The 7th constellation from Janma Nakshatra is called Naidhana Nakshatra. Naidhana means "death" and this Nakshatra shows death and suffering. (7) The 12th constellation from Janma Nakshatra is called Desa Nakshatra. Desa means "country" and this Nakshatra shows one's country. (8) The 13th constellation from Janma Nakshatra is called Abhisheka Nakshatra. Abhisheka means "coronation" and this Nakshatra shows power and authority. This is also called Raajya Nakshatra (kingdom). (9) The 19th constellation from Janma Nakshatra is called Aadhaana Nakshatra. Aadhaana means "epoch/conception" and this Nakshatra shows well-being of family. (10) The 22th constellation from Janma Nakshatra is called Vainaasika/Vinaasana Nakshatra. Vainaasika means "destructive" and this Nakshatra shows one's destruction. (11) The 25th constellation from Janma Nakshatra is called Maanasa Nakshatra. Maanasa means "mind" and this Nakshatra shows one's mental state.

2.3 Synastry by Nakshatra

2.3.1 Aswini

General: He/She is daring, handsome, truthful, prosperous, self-sufficient, intelligent, popular, knowledgeable, rich, well mannered, expert, outspoken, explorer, sportsmanship, and has vitality, courage, dynamism, initiative, action and a contented family life.

1. Physical features: Male has a beautiful countenance, bright and large eyes, broad forehead and big nose. Female eyes will be bright, but small like a fish with a magnetic look.

2. Character: He will remain faithful and will not hesitate to sacrifice anything for beloved persons. He keeps his patience even at the time of greatest perils. He is the best advisor to the persons in agony. He is a firm believer of God. He keeps the entire surroundings neat and clean. Female attracts people with her sweet speech. She maintains utmost patience. She indulges in too many sexual operations. She respects elders.

3. Education and sources of earnings / profession: Male will be full of struggle up to his 30th years of age. There will be steady and continuous progress after 30 years and will continue up to 55 years of age. He meets his desires and needs at any cost. Female **will do** administrative job in office. She may quit the job or seek voluntary retirement after 50 years of age.

4. Family life: Male loves his family with sincerity. His own family members will hate him due to his adamant behaviour. He cannot derive affection and care from his father. In other words, his father will neglect him as also his co-born. Maximum possible help will come from outside his family circle. Normally marriage takes place between 26 to 30 years of age. There will be more sons than daughters to him. Female's marriage takes place at a young age i.e., before the age of 23 years and marriage ends either with divorce or separation or even death of the husband. There will be more female children than male children. She responds to the needs and desires of her children. She will be busy in the welfare of the family and society.

5. Health: Health is good. She/He will have mental worry and anxiety and may face brain disorder of mild type at a later stage.

Positive traits: They have courage, daringness and swiftness of action, action oriented nature, intellectual ability and acute sharpness.

Negative traits: They give the rash impulsiveness with unnecessary risks and blunders creating danger and troubles and have revengefulness and stubbornness in nature.

Career: They excel as physicians, entrepreneurs and in areas of adventure sports, military or armed forces, law enforcements and athletic. He/She does business related to machinery, iron, steel, mineral, writing, publishing and engineering. Some has officer grade posting in architecture, stock broking, police, army, interior designer and flying, driving, riding and sports field. He/She is predisposed to muscular injuries and should be careful, as accidents are common.

1st Partner Nakshatra	2nd Partner Nakshatra	Match / Synastry Status
Aswini (Mesha Rasi)	Bharani - Aridra - Pushya - Anuradha - Purva Ashadha - Sharvana Shathabisha - U. Bhadrapada	Uttham
Aswini (Mesha Rasi)	P. Bhadrapada - Dhanishta - Uttara Ashadha - Visakha - Purva Phalguni - Punarvasu – Mrigashira (3&4th Pada) - Chitra - Rohini – Krittika (1st Pada)	Madhyam

3.4.2 Bharani

General: He/She is stern disciplinarian, a ruthless, truthful, law-abider, intelligent, healthy, very courageous, long lived, very creative and does painting and photography. He/She dislikes being controlled and manipulated. He/She may have problems in romance and prone to putting on weight and never shies away from a fight. Obesity and skin diseases could plague him/her. He/She has a possibility of getting hurt near head and eyes. He/She is excessively indulged sexually. They are long lived. He/She desires to attract opposite sex and is marked by large expressive eyes and mesmerizing smile.

1. Physical features: Male is of medium size, and has less hair, large forehead, bright eyes and beautiful teeth. Female will have a very beautiful figure with white teeth.

2. Character: Male does not like to harm or buttering anybody. He has to face a lot of resistance and failures. He will spoil the relationship even with near and dear ones. When the opponent comes forward with folded hands, he will completely forget the enmity. His arrogance and subordination is equal to death for him and will lead into trouble. Female possesses a clean, admirable and modest character. She respects her parents and elderly persons. She is bold, impulsive and over optimistic.

3. Education, sources of earnings / profession: Male will have a positive change in his life and livelihood after his 33 years of age. He can be successful in the business of tobacco items or cultivation of tobacco. Female earns her own livelihood and will be successful as a receptionist, guide or sales woman or in sports activities.

4. Family life: Male will get married around 27 years of age. He is fortunate in conjugal bliss. His spouse will be expert in household administration and well behaved. His father will die, if his birth is in the 1st or 2nd quarter of Bharani. Female gets married around her 23rd years of age. She will have an upper hand and will behave like a commander. She will enjoy full confidence, love and satisfaction from her husband but her in-laws will trouble her. She will over power her husband in all walks of life, but is not aggressive.

5. Health: Male is a very poor eater. There is chance of injury in the forehead and just around the eyes. He is a chain smoker. Female will have frequent menstrual problems, uterus disorders, anaemia and in some cases tuberculosis has also been noticed.

Positive traits: They have poise, calmness, multiplicity of interests, sincerity and dutifulness towards their liabilities.

Negative trait: They are indulge in material and physical pleasures, vulnerable to luxuriousness and extravagance, unable to fit up with any form of domination and source of authority and has disregard for others' sentiments.

Career: His/Her careers will be in military, chemical industry, medicine or agriculture. He/She may be a social reformer, activists and philosophers. They excel in arts and curve a niche in careers based on singing, dancing, acting and painting. They also make excellent administrators, businessman and ingenuous surgeons and adjudicators.

1st Partner Nakshatra	2nd Partner Nakshatra	Match / Synastry Status
Bharani (Mesha Rasi)	Aswini - Krittika 1st - Mrigashira(3&4th Pada) - Punarvasu - Ashlesha Chitra(3&4th Pada) - Visakha- Jyeshtha - Moola - Uttara Ashadha - Revati	Uttham
Bharani (Mesha Rasi)	Shathabisha - Sharvana - Swati - Aridra Krittika (2,3&4th Pada) Magha – Visakha (4th Pada)	Madhyam

3.4.3   Krittika

General: He/She is brilliant, very prosperous, learned, powerfully built and possesses wealth. He/She has a magnetic & commanding personality and popularity in the opposite sex, leadership skill, penetrating ability for success. He/She will be good at carpentry, sculpture, metal work, and interior decoration. He/She runs after other women and is heavy eaters and fiery. He/She also gets into bad habits like gambling, drinking and arguing. He/She is stubborn, aggressive, and very angry. He/She is thin and good cook. He/She has frequent illicit sexual affairs. He/She is endowed with son and enjoys life.

1. Physical features: Male is solidly built body with big shoulders and developed muscles. He has a commanding appearance. Female has extremely beautiful body, medium height. .

Character: Male does not like to achieve name, fame and wealth through unfair means or at the mercy of others. His need is less and does not believe in age-old blind belief and customs. He does public work but looses and fails. Frustrations lead him to outburst. Female will not subdue to the pressure of others. Therefore, she has to suffer mentally. She is fond of quarrels. She expresses arrogance not only in her appearance but also in the domestic environment.

3. Education, source of earning/profession: Male will earn his livelihood in the foreign land or away from the place of birth. Partnership business is not suitable to him. In case he is interested in business, he can derive maximum benefits from the yarn export, medicines and decorative industries. Female is not highly educated. She earns by working in the paddy field, agriculture field, fisheries and other small-scale industries. She also earns as a musician, artist, tailor or leather products manufacturer.

4. Family life: Male is lucky in his Love marriage with a faithful and virtuous wife. The health of the spouse will be a concern for him and he may have to quite often live separately due to work or due to ill health of the spouse. He is more attached to mother and will enjoy more favour and love from the mother. His father will be a pious man and well-known person; he cannot enjoy comforts and benefits from father. The period between 25 to 35 years and 50 to 56 years will be very good. **Female** cannot enjoy full comforts of her husband. In some cases childlessness or separation from husband or no marriage is indicated. She will not keep cordial relation with relatives. She lives in an illusion world and she shows discourtesy to the people who actually are her well wishes.

5. Health: Male is prone to dental problem, weak eyesight, tuberculosis, wind and piles, brain fever, accident, wounds, malaria or cerebral meningitis. Female health will be severely afflicted either due to excess work or lack of mental peace. Glandular tuberculosis has also been noticed.

Positive traits: They have honesty, frankness and respect of people around, protective ability, and are extremely hard working.

Negative traits: They have tendency of fault finding, anger, lack of diplomacy and are centre of negative criticism on account of their bluntness.

Career: He prospers as administrators, leaders, lawyers and excels in career away from their home land. The partnership will never be his cup of tea, but can excel in business related to yarn, artistic goods and medicines, engineering and draftsman ship.

Ideal Partner by Nakshatra: This table gives you the ideal matchmaking and Compatibility between between Partners for running a Business.

1st Partner Nakshatra	2nd Partner Nakshatra	Match / Synastry Status
Krittika 1st Pada (Mesha Rasi)	Aswini - Bharani - Aridra - Pushya - Hasta -Swati - Anuradha - Moola - Shathabisha - U. Bhadrapada	Utthamam
Krittika 1st Pada (Mesha Rasi)	Mrigasira (3&4th Pada) - Magha - Chitra - Jyeshtha - Dhanishta - Revati	Madhyam
Krittika 2,3,4th Pada (Vrishabha Rasi)	Aswini - Bharani - Pushya - Magha - Swati - Anuradha - Moola - Shathabisha - U. Bhadrapada	Uttham
Krittika 2,3,4th Pada (Vrishabha Rasi)	Revati - Dhanishta - Jyeshtha - Hasta - Purva Phalguni - Rohini	Madhyam

2.4.4   Rohini

General: He/She is long-lived, religious, experts, well-behaved, master in conversation, genius, popular, wealthy and has excellent persuasive skills, devotee of parents and kind hearted. His/Her work areas are business. He/She can be over sexual. He/She can rise to the top and achieve his/her desires and has children. He/She respects Gods and Brahmins. He/She is highly imaginative and love to bask in the fictional world. His mind is restless like a deer, and that is the reason why she cannot stick to a single thing at one time. He/She is the lovers of art, music, dance and sculpture. He/She is good photographers and takes active part in any type of art forms. He/She has a very sweet and loving personality.

1. Physical Features: Male is slim in physique; normally dark, short structured and has very attractive eyes with a special magnetic touch. Appearance is very beautiful and attractive, big shoulder and well developed muscles. Female is beautiful. Her eyes are very attractive. She is of medium height and fair complexion.

2. Character: Male has special knack to find faults of others and his brain rules him. He spends everything for today's comfort. He lacks patience and forgiveness. He fails to pass through the path of requisite system and has the downfall. He blindly believes others. Female is well behaved and well dressed with a lot of pomp and show. She has a very weak heart, but is short tempered and invites troubles.

3. Education, sources of earning / profession: Male may earn from business of milk products, sugar cane or engineering goods. He is best adapted for mechanical or laborious work. He faces a lot of problems economically, socially and on health grounds during the age of 18 to 36 years. He will enjoy life best between the age of 38 to 50 years and 65 to 75 years of age. Female earns from oils, milk, hotels, and paddy fields and as a dressmaker. Middle level education is indicated for her.

4. Family life: Male cannot enjoy full benefit from his father and is more attached to his mother and maternal uncle. His married life will be marred with disturbances. Female family life is good with a full comfort from her husband and children. She must avoid doubting her own husband as otherwise marriage may end in divorce.

5. Health: Male is prone to diseases connected with blood cancer, jaundice, urinary disorders, blood sugar, tuberculosis, respiratory problems, paralysis and throat trouble. Female health is good. She will have pain in the legs, in the breast, and may have breast cancer, irregular menses and sore throat, pimples and swelling above the neck.

Positive traits: He has temperamental sweetness and gentleness and they are so fine tuned that they seldom get into fits of temper. Their softly affectionate nature does not allow offence or revenge and a blend of perseverance.

Negative traits: They have lack of purpose, stability and imagination; lack of endurance and patience; struggle, indecision and fickle mindedness.

Career: Natives are efficient sculptors, artistes, musicians, danseuse and creative directors. They can make mileage out of career options including photography, editing, agriculture, environment, advertisement, writing, marketing and jewellery designing.

Ideal Partner by Nakshatra: This table gives you the ideal matchmaking and Compatibility between a Girl and a Boy for Marriage or between Partners for running a Business.

1st Partner Nakshatra	2nd Partner Nakshatra	Match / Synastry Status
Rohini (Vrishabha Rasi)	Bharani - Krittika - Mrigashira - Punarvasu (4th Pada) - Ashlesha – Uttara Phalguni (1st Pada) – Chitra (3&4th Pada)- Visakha - Jyeshtha - Uttara Ashadha - Dhanishta - P. Bhadrapada - Revati	Uttham
Rohini (Vrishabha Rasi)	U. Bhadrapada - Anuradha - Pushya - - Punarvasu (1,2&3rd Pada - Aswini	Madhyam

2.4.5   Mrigashira

General: He/She is truthful, handsome; enjoy wealth and prosperity, pure in heart, knowledgeable in architecture, well at administration, loved by the king, full of energy, very popular and very creative. He/She travels a lot. He/She is successful research fellows and highly successful in conducting investigations, highly intelligent, and has beautiful children and many sexual relationships. He/She is restless, gentle, peaceful, and always travelling.

1. Physical feature: Male will have a beautiful and stout body, tall, moderate complexion, thin legs and long arms. Female is tall, sharp look, leaning body. Her countenance and body are very beautiful.

2. Character: He is very sincere in his dealings. He may get cheated in partnership business. He puts blind trust on others and gets frustration and repentance. He may be ditched. He is an inborn coward. He will not have peace of mind and get irritated even on small matters. His life will be painful up to 32 years of age. His life starts settling down to the maximum satisfaction after that. Female takes keen interest in the social work. She is quick-witted and selfish. She has a poisonous tongue, but is educated and fond of fine arts. She acquires considerable wealth and enjoys good food. She will be lucky and have ornaments and fine clothing. She is greedy for wealth.

3. Education, sources of earning / profession: Male will have good education. He is a very good financial adviser. He achieves success in business after the age of 32 years. He achieves the benefits unexpectedly during the good period at 33 to 50 years of age, but it will be wasted later on due to his own fault. Female will attain good knowledge in the mechanical or electrical engineering, telephones, electronics. She is more interested in the jobs that are normally done by males.

4. Family life: Male will not derived benefit from the co-born. Co-born will cause troubles and problems to him and will maintain extreme enmity and harm him. His spouse is attracting romantic, may not keep good health. Moreover, his married life is not cordial due to his adamant nature. Female keeps her husband under her control. She may have one or two love affairs in her early life, which will not culminate into marriage. However, after marriage she is very much attached to her husband, as if, nothing has taken place in the past. She will have children and is devoted to her husband.

5. Health: Male may face ill health during his childhood, like constipation ultimately leading to stomach disorder, cuts and injuries, pains in the shoulders near collarbones. Female is prone to goitre, pimples, venereal diseases, menstrual troubles and shoulder pains.

Positive traits: He/She is truthful, clean at heart, intelligent and has good administration powers and are obedient, respecting his teachers and always keen to learn and observe.

Negative trait: He/She is restless and nervous, impatient, takes wrong decisions and commits mistakes.

Career: He/She is musician, tailor, engineer, communicator or traveller. Careers are sports, advertising, communication, and environment campaigns, travel related industry.

Ideal Partner by Nakshatra: This table gives you the ideal matchmaking and Compatibility between a Girl and a Boy for Marriage or between Partners for running a Business.

1st Partner Nakshatra	2nd Partner Nakshatra	Match / Synastry Status
Mrigashira 1 & 2nd Pada (Vrishabha Rasi)	Aswini - Krittika - Rohini - Pushya - Uttara Phalguni (1st Pada) - Anuradha - Moola – Uttara Ashadha (2,3&4th Pada) - Sharvana - Shathabisha - U. Bhadrapada	Uttham
Mrigashira 1 & 2nd Pada (Vrishabha Rasi)	Revati - P. Bhadrapada - Jyeshtha - Visakha - Ashlesha - Swati – Punarvasu (4th Pada) - Bharani - Purva Ashadha	Madhyam
Mrigashira 3 & 4th Pada (Mithuna Rasi)	Aswini - Krittika - Rohini - Aridra - Uttara Phalguni - Hasta - Anuradha - Moola - Uttara Ashadha (1st Pada) - Sharvana - Shathabisha - U. Bhadrapada	Uttham
Mrigashira 3 & 4th Pada (Mithuna Rasi)	Revati - P. Bhadrapada – Jyeshtha - Swati - Visakha - Pushya - Purva Ashadha – Punarvasu (1,2&3th Pada)- Bharani	Madhyam

2.4.6   Aridra

General: He/She is hard working, proud, cruel, lower levels person, bereft of money, expert in trade and commerce, and does jobs which are forbidden. He/She life undergoes many ups and downs and U-turns and faces many challenges in his life. He/She is little educated, long lived, and little interested in doing things and more inclined towards sex. He/She causes pain to her loved one and suffers due to hunger and hates all. His/Her violent temperament causes many tears and depression to him and leads to his/her early destruction and death.

1. Physical features: Male has different shape and structure and is slim. Female has handsome body with charming eyes, prominent nose.

2. Character: Male is good psychologist. His dealings with his friends and relatives will be of very much cordial type. He does not have a constant type of behaviour Female is well behaved and peace minded. She is intelligent, helpful to others and clever in finding fault in others. Some of them may have two mothers or two fathers.

3. Education, sources of earning/profession: Male is over sincere in the work or his business. He is selfless social workers and earns his livelihood away from his home and family. He is settled in foreign places. He has golden period between 32 to 42 years of age. Career best suited is lower level in fire fighting, farming, gardening, writing, postal services, transport, education, police and defence. Female will attain distinction in the educational or scientific field. She may specialize in electronic or pharmaceutical. She also has consultation works.

4. Family life: Male marriage will be delayed. In case marriage takes place early he will be compelled to live separately from the family due to difference of opinion. When the marriage takes place at a late stage his married life will be good. His spouse will exercise full control over him. Female marries at a late stage. She enjoys love and affection from her husband and husband's family. Her married life will be full of thorns. Even children cannot give her the required happiness. In some cases, there will be either death of the husband or divorce takes place.

5. Health: Male may have some incurable diseases like paralysis, heart trouble and dental problems. He is also prone to Asthma, esonophilia, and dry cough, ear trouble. Female is prone to menstrual troubles, asthma, spoiled blood, lack of blood or uterus trouble, ear trouble, and mumps, bilious and phlegmatic.

Positive traits: He/She is hard working, intelligence. He/She renovates old things and produce new things with their fertile thinking and so they are strong and mentally stable and at the same time empathetic towards others.

Negative traits: He/She is always sad and a kind of impulsive mentality, cleaver and calculating, stubborn and may exhibit violent temperament, which cannot help them to make very close social bonds.

Career: His career lies in fields like transport industry, communication industry, financial broking firms, shipping industries.

Ideal Partner by Nakshatra: This table gives you the ideal matchmaking and Compatibility between a Girl and a Boy for Marriage or between Partners for running a Business.

1st Partner Nakshatra	2nd Partner Nakshatra	Match / Synastry Status
Aridra (Mithuna Rasi)	Bharani - Mrigashira – Punarvasu (1,2&3rd)- Purva Phalguni – Chitra(1&2nd Pada) - Purva Ashadha - Dhanishta -Vishakha (4th) - P. Bhadrapada - Revati	Uttham
Aridra (Mithuna Rasi)	U. Bhadrapada - Uttara Ashadha - Moola - Uttara Phalguni - Magha Punarvasu (4th Pada) - Krittika - Aswini	Madhyam

2.4.7 Punarvasu

General: He/She is endowed with children, long lived, loved by wife, and possesses wealth. He/She has utmost happiness. He/She will share his material possessions with others and will choose a religious or spiritual career. His/Her deep-rooted faith in God and high morals will give his/her family a deep sense of security.

1. Physical features: Male is handsome and has long thighs and long face. There will be some identification mark on the face or on the backside of the head. **Female** eyes are red and have curly hair, sweet speech and high nose.

2. Character: Male does not like to cause trouble to others. He will lead a simple life. **Female** has argumentative tongue, which leads her into frequent friction with her relatives and her neighbours. She will have many servants and a comfortable life.

3. Education, sources of earning/profession: Male is in business, journalism, publishing, auditing and writing and is contended with little and sticks to ancient tradition and belief. He attains much name and fame as a teacher or as an actor and physician. He will not do so well up to 32 years of age. He will not accumulate wealth but attain public honour. He lacks business trick and is straightforward. He is an innocent and frustrated looking. **Female** gets mastery over dances.

4. Family life: Male will have very happy married life. Children feel loved and secured. He is the most obedient child of his parents. He respects his father and mother as also his teacher. His married life may not be good. He may either divorce his wife or get involved in another marriage. In case he does not go for the second marriage the health of the spouse will give a lot of problems and mental agony. However, his spouse will have all the qualities of a good housewife. **Female's** husband will be most handsome man.

5. Health: Male has diseases related to lungs, blood, waist, and ear and dental. There is not any serious health problem. He drinks lot of water. He has strong digestion. **Female** cannot enjoy good health and has Jaundice, tuberculosis, goitre and pneumonia, stomach upset and ear trouble.

Positive traits: He/She is spiritual, sober, pleasant, social, and has tendency to discern a spiritual meaning of life, essential domestic or home spun nature.

Negative traits: He/She has lack of forethought, carefulness, susceptibility, and disregard for people, fickle mind and instability.

Career: He/She excels as spiritual teachers, psychologists and mystic philosophers, architecture, innovations, civil engineering and maintenance of buildings. Acting, dramatics and writing are also appealing for them.

Ideal Partner by Nakshatra: This table gives you the ideal matchmaking and Compatibility between a Girl and a Boy for Marriage or between Partners for running a Business.

1st Partner Nakshatra	2nd Partner Nakshatra	Match / Synastry Status
Punarvasu	Aswini - Mrigashira - Aridra	Utthamam

1, 2, 3rd Pada (Mithuna Rasi)	- Pushya - Chitra (1&2nd Pada) - Anuradha - Moola - Dhanishta - Shathabisha - U. Bhadrapada	
Punarvasu 1, 2, 3rd Pada (Mithuna Rasi)	Revati - Sharvana - Purva Ashadha - Jyeshtha - Hasta - Pushya - Ashlesha – Chitra (3&4th Pada - Swati - Rohini	Madhyam
Punarvasu 4th Pada (Karaka Rasi)	Aswini - Mrigashira - Aridra - Pushya Chitra - Swati - Anuradha - Moola - Dhanishta - U. Bhadrapada - Shathabisha	Uttham
Punarvasu 4th Pada (Karaka Rasi)	Revati - Sharvana - Jyeshtha - Purva Ashadha - Hasta - Ashlesha - Rohini - Bharani	Madhyam

2.4.8   Pushya

General: He/She is fortunate, intelligent, self-sufficient, helping hand, important person in the world, knowledgeable, respected by kings, wealthy, and obeys his parents and possesses wealth and vehicles of novelty. He/She is caring, nurturing, kind, and helpful. Pushya is one of the most lovable Nakshatra. He/She is honest. He/She believes in the rules, and forces others to follow the laws. He/She is teachers, preachers and professors. He/She could be predisposed to ailments like gall stones, gastric ulcers, skin diseases.

1. Physical Features: Male will have a scar or a black mole on his face. **Female** has a short stature, moderate complexion, well-proportioned. Generally, he will be beautiful.

2. Character: Male is very weak by heart and doesn't reach at conclusion on any matter. His outward expression is hypocritical against his consciousness i.e. the inner expression is negative and the outward expression is positive. He is fond of good dresses. **Female** is peace-minded, very submissive to the elders but oppressed by all. She is sincere, affectionate but moody. She is religious and god fearing. She respects her elders. She is conservative. She does systematically and methodical action.

3. Education, sources of earning/profession: Male is employed in state government jobs like medicine, social work, health care. He will have utter failure. He has several obstacles because of no basic education. He does work with utmost sincerity and certainty. His success is as probable. He faces the grip of poverty up to the age of 15-16 years of age and thereafter he will enjoy a mixture of good and bad up to the age of 32 years. There will be remarkable all round improvements i.e. economically, socially and in health from the age of 33 years. He will stay away from his native place and family for his bread earning. He undertakes the work where maximum travelling is involved. Female may have income from Agriculture, land and buildings. She may be employed in a job where maximum trust and secrecy is required e.g. private secretaries, secret departments of a country.

4. Family life: Male will have lot of problems in the family circle and he will have to depend on others for even day-to-day requirements. He will be forced to stay away most of the time from his family. He is very much attached to his parents. **Female** is devotee and attached to her husband only, but her husband quite often mistakes her for her moral character. She will have good duty bound children.

5. Health: Male will not keep good health during his childhood up to the age of 15 years. Gallstones, gastric ulcer, jaundice, cough, eczema and cancer are the possible diseases. Female will have respiratory problem. She is prone to tuberculosis, ulcers, breast cancer, and jaundice, eczema, bruises in the breast or gastric ulcer.

Positive traits: He/She is benevolent, philanthropy and humanitarian, cool and has good manners, protection against destruction and violence, inherent qualities of creation and expansion.

Negative traits: He/She is zealously protective of his family, society or community to which they belong, orthodox, narrow-minded and possessive, conservative and prejudice.

Career: He/She excels in careers related to counselling, public administration, planning, research, good priests or clergies. He/She is geologist, developer of land, aquatic biologist, land or agricultural merchants, philosophers, religious leaders, teachers and professors.

Ideal Partner by Nakshatra: This table gives you the ideal matchmaking and Compatibility between a Girl and a Boy for Marriage or between Partners for running a Business.

1st Partner Nakshatra	2nd Partner Nakshatra	Match / Synastry Status
Pushya (Karaka Rasi)	Rohini - Aridra - Punarvasu - Ashlesha - Hasta - Swati - Vishakha - P. Bhadrapada - Revati Shathabisha	Uttham
	Aswini - Kritika - Mrigashira - Uttara Phalguni - Chitra - Moola - Uttara Ashadha (2,3&4th Pada) - Dhanishta - Magha	Madhyam

2.4.9 Ashlesha

General: He/She is born leaders, daring, angry, wicked, and harmful to people; and suffers on account of sex starvation, pay fines by doing sinful acts and causes anguish to others. He/She spends his wealth for evil purposes, and losses on investments. His/Her work areas will be business, contracts or teaching. He/She can hypnotize any one with glare of eyes and uses it for black magic. He/She is cold blooded dangerous man. He/She is extremely sexual. He/She is always sad, vary bad person, highly stingy, and develops enmity with his own people and others. (i) If born in the first pad, he/she is impotent. (ii) If born in the second pad, he/she is a servant always serving others. (iii) If born in the third pad, he/she suffers from diseases. And (iv) if born in the fourth pad, him /her, though fortunate, has medium life.

1. Physical Features: Male thinks and walks fast. He is normally dark, possesses rude appearance and rude features but inside built up is void. Female is not good looking. However, she will have beautiful figure.

2. Character: Male is grateful to anybody, very talkative; shines in the political field and can give suitable leadership to a country. He does not believe others blindly. He mostly associates himself with black marketers, thieves and murderers. He does not keep any distinction between rich and poor, good and bad people. He always supports the persons who are not in need and reject the request of the persons who are in need. He is luckiest, popular but hot-tempered person.

Female is shy. Her moral is of a very high order. She will enjoy good respect and recognition from her relatives. She conquers her enemies.

3. Education, source of income/profession: Male has education in Arts or Commerce and changes careers many times. There will be a heavy loss of money at the age of 35-36th years and unexpected and unearned income at 40 years of age. Female does official work. She may be employed in an administrative capacity. Illiterate females may be engaged in selling fishes or work in the agricultural field.

4. Family life: Male shoulders family responsibilities but his wife does not want to share her wealth with other relatives such as sister-in-law or cousin brother Female is very efficient in the household administration. Her in-laws can make some plot against her so as to create friction with her husband.

5. Health: Male will have flatulence, jaundice, and pains in legs and knees, stomach problem. He will be addicted to drugs. Female will have frequent nervous breakdown, Joint pains, hysteria and jaundice.

Positive traits: He/She has intuition, overcoming a situation steeped in danger, planning, cool calculation, and courage and leadership disposition.

Negative traits: He/She displays cold ruthlessness, deceit and an aura of suspicion and is endowed with unattractive features and has suspicion and cunningness and merciless guile, insidious harmfulness, dependence on others and shamelessly plotting nature, deception, dishonesty and falsehood, miserliness, selfishness, ungratefulness, acute suspicion, depression, anxiety, schizophrenia and total absence of benevolence.

Career: He/She excels in careers related to cunning politicians, filing law suits, lawyers, advocates, and business men. He/She also does well in entertainment industry, occult mysticism and astrology. With the planetary lordship of Mercury, he/she succeed more in business than in profession.

Ideal Partner by Nakshatra: This table gives you the ideal matchmaking and Compatibility between a Girl and a Boy for Marriage or between Partners for running a Business.

1st Partner Nakshatra	2nd Partner Nakshatra	Match / Synastry Status

Ashlesha (Karaka Rasi)	Kritika - Mrigashira - Punarvasu – Pushya – Chitra – Visakha(1,2&3rd Pada) - Anuradha – Dhanishta – P. Bhadrapada – U. Bhadrapada	Uttham
Ashlesha (Karaka Rasi)	Bharani – Rohini – Aridra – Hasta – Uttara Phalguni (2,3&4th Pada)- Uttara Ashadha – Sharvana - Shathabisha	Madhyam

2.4.10 Magha

General: He/She is highly sexy, learned, honest, helpful, devotee of parents and destroys enemies. He/She has royal and respectable positions and authoritative status in his life. He/She is leader and has material pleasures of the world. He/She is enterprising nature, thoughtful interaction with other. He/She respects his father and gets to the top. He/She will enjoy a harmonious marital life. He/She may do well in business and can amass a lot of wealth. He/She is a brilliant planner and organiser and is always in command of a situation. He/She may take top most charge and has a drive for power and wealth like a C E O. He/She is long lived, served by many servants, helps their relatives.

1. Physical features: Male has a prominent neck and a hairy body, a mole in the hands and also beneath the shoulder and is medium size height and innocent looking countenance.
Female has a most beautiful and attractive feature. If Saturn aspects Moon in this Nakshatra, she will have long bunch of hair.

2. Character: Male is God fearing, soft spoken and leads a noiseless life. He receives honour and recognition from the learned persons. He is hot-tempered and cannot tolerate any action or activity, which is not within the purview of truth. He/She is charitable and God fearing. She will enjoy royal comforts. She can attain mastery over both household and official activities. She helps without selfish motive.

3. Education, sources of earning/profession: Male attains much success due to his straightforwardness. His dealings with his superiors and subordinates are very cordial and most technical. Hence, he is a link between his bosses and subordinates. **Female** is employed in managerial position like a queen if Jupiter is placed in this Nakshatra. She is married to a wealthy person.

4. Family life: Male will enjoy a happy and harmonious married life. He shoulders several responsibilities and burden of his co-born. Female invites friction in the family; makes conflict between her husband and in-laws, resulting in mental torture for everybody in the family. She will have good children; preferably the first will be a son and next two daughters.

5. Health: Male may suffer from night blindness, cancer, asthma or epilepsy if Saturn and Mars jointly aspects Moon in this Nakshatra or conjoin in this Nakshatra. Female's eyes may be affected, and has blood disorders, uterine trouble and hysteria.

Positive traits: He/She is religious and traditional in nature, god fearing and respects his elders, deeply follows his forefather's teachings and abides by that strictly. His empathetic behaviour, care towards other's feelings and possessing a careful approach towards life is another positive traits.

Negative trait: He/She is often attracted to material pleasures and is proudly due to enjoyment of high positions in the society. He/She is often over confident being in powerful position, and imposes his decisions on others without much consideration, which causes resentment in some people about him.

Career: He/She works hard and enjoys a very good position. His career interest is to work independently as bosses, leaders and switch his jobs frequently to achieve the position he desires.

Ideal Partner by Nakshatra: This table gives you the ideal matchmaking and Compatibility between a Girl and a Boy for Marriage or between Partners for running a Business.

1st Partner Nakshatra	2nd Partner Nakshatra	Match / Synastry Status
Magha (Simha Rasi)	Bharani – Aridra – Pushya – Swati – Anuradha – Sharvana – Shathabisha – U.	Uttham

| | Bhadrapada | |
| Magha (Simha Rasi) | Kritika – Purva Phalguni – Chitra (3&4th Pada) – Hasta - Dhanishta – P. Bhadrapada | Madhyam |

2.4.11   Purva Phalguni

General: He/She has sensual pleasure; prosperity, enjoyment, good fortune and luck. He develops keenness towards arts from childhood and is an artist. He/She is experts in love making, wealthy, long lived and loves his/her brothers, gets few children. He/She may incur a lot of debt. His/Her marital life will be happy. He/She is devotee of parents, religious, good speaker and generous. He/She is very sexual and passionate. He is peace loving and keep distance from disputes, and tries to bring in an amicable solution to the problems without fighting. He is greatly liked and loved by people around them.

1. Physical features: He has attractive personality with a stout body and a snubbed nose. She will be medium height and has overall an attractive personality.

2. Character and general events: He has intuitive powers to know others. He extends his helping hand to the needy. He has sweet speech. He is fond of travelling. She is polite, artist and does charitable deeds. She has self-imposed showy image and believes that there is none above her.

3. Education, sources of earning/profession: He cannot take up a job of subordination. He cannot be a 'yes master' and so cannot derive much benefit from superiors. He does not like to get any benefit at the cost of others pocket. He is able to crash his enemies and attain much success in all the work he undertakes. He gives preference to the position rather than money. He frequently changes job. The borrowers will not return the money to him. He will reach a good position where power and authority vests in him after his 45th years of age. In the business field he can shine well. She attains a good degree of education of scientific subjects. She may be a lecturer. She will have reasonable wealth.

4. Family life: His married life will be happy. He will have good wife, children and derives much happiness from them. He leads life away from native place and family members. She has a loving husband and children. She is duty bound wife. Her daredevil attitude creates friction between the family members. She cannot have a cordial relation with her neighbours due to her arrogance nature and position of her husband. His health is good and no permanent nature of disease is noticed in the native.

5. Health: She suffers from frequent menstrual, asthma, jaundice or breathing trouble.

Positive traits: He maintains a cordial and warm relationship with others and relatives. He is very loving and affectionate and he cares for the people close to him and helps them to the utmost when it is needed. He keeps homes and surroundings clean and decorates them with beautiful things.

Negative traits: He is over confident and arrogant about his appearance and deeds. He is restless and loses patience easily.

Career: He studies in areas like criminal psychologies, research and administrative studies. The suitable career areas for him are government services, transport management, automobile industries and hotel industries. He/She will do well in the legal or law enforcement field.

Ideal Partner by Nakshatra: This table gives you the ideal matchmaking and Compatibility between a Girl and a Boy for Marriage or between Partners for running a Business.

1st Partner Nakshatra	2nd Partner Nakshatra	Match / Synastry Status
Purva Phalguni (Simha Rasi)	Aswini – Kritika - Aridra - Magha – Uthiram1 – Chithirai3&4 – Visakha – Jyeshtha – Uttara Ashadha (2.3&4th Pada) - Dhanishta – P. Bhadrapada – Revati	Uttham
Purva Phalguni (Simha Rasi	Aridra - Swati - Moola - Sharvana - Shathabisha	Madhyam

2.4.12   Uttara Phalguni

General: He has kind heart, charity, honesty, truthfulness, patronage and knowledge. He has wisdom in science and arts and keeps away from disputes and quarrels. He/She is wealthy, famous, kingly, hard working, impotent, and achieves fame. He/She earns well out of commissions and royalties. Marital life is pleasant and contented. He/She is business minded and has an association with bad people. He/She is courageous, good with people, leading him/her to powerful positions.

Male Natives

1. Physical features: Normally, he is a tall and fat figure, large countenance and long nose and a black mole in the right side of his neck.

2. Character: He is extreme sincere and good in social work. He does not have patience or tolerance. However, he will not admit his fault at any cost.

3. Education, sources of earnings / profession: He shines in the profession of public contact and earns a good amount as commission out of the public dealings. He is hard-working and reaches a good position. He is suited to the profession as a teacher, writer or research. He earns extra money out of tuitions. There will be complete darkness up to 32 years of age. Thereafter, his progress is much faster and achieves much of his desires from his 38th years of age up to his 62nd years of age. He will earn fame and wealth during his fifties. He has self-acquired assets. He is good in mathematics or engineering.

4. Family life: His married life is good. He is contended and wife will be most efficient.

5. Health: His health will be generally good. He is prone to dental problems, gastric trouble, liver and intestine problems.

Female Natives

1. Physical features: She has medium height and a black mole on the face.

2. Character and general events: She will have very calm and simple nature. She does not keep enmity for long. She is principled, always joyful.

3. Education, sources of earnings / profession: She has mathematical aptitude or scientific background. She is a teacher or lecturer or in administrator in sanitary department or hospitals. She may also earn money as a model or as an actress.

4. Family life: She has happy family life. She is wealthy and very clever in managing domestic work. Neighbours may create problems in family life in alliance with her in-laws.

5. Health: She enjoys good health with diseases like Asthma, menstrual and headache.

Positive traits: He has intrinsic greatness of head and heart, determination, conviction, esteem, knowledge and passion for the delicate subtleties of life.

Negative traits: He has extra marital affairs, obsession with cleanliness and arrogance and he can be inconsiderate and disdainful.

Career: He does well in careers related to scientific research, astronomy, media, sales, philanthropy, acting, writing, armed forces and military.

Ideal Partner by Nakshatra: This table gives you the ideal matchmaking and Compatibility between a Girl and a Boy for Marriage or between Partners for running a Business.

1st Partner Nakshatra	2nd Partner Nakshatra	Match / Synastry Status
Uttara Phalguni 1st Pada (Simha Rasi)	Aswini – Bharani – Rohini – Aridra – Pushya – Magha – Purva Phalguni - Swati – Anuradha – Sharvana – Shathabisha – U. Bhadrapada	Uttham
Uttara Phalguni 1st Pada (Simha Rasi)	Revati – Dhanishta – Jyeshtha - Ashlesha – Mrigashira - Purva Ashadha - Moola	Madhyam
Uttara Phalguni 2, 3 & 4th Pada (Kanya Rasi)	Aswini – Bharani – Rohini – Aridra – Pushya – Magha – Purva Phalguni - Hasta – Anuradha – Moola – Purva Ashadha – Shathabisha – U. Bhadrapada	Uttham
Uttara Phalguni 2, 3 & 4th Pada (Kanya Rasi)	Revati – Dhanishta (3&4th Pada) – Jyeshtha – Swati - Ashlesha – Mrigashira	Madhyam

2.4.13 Hasta

General: He has inclination towards music. He/She is warrior, religious minded, learned, wealthy, daring, helpful to others, God fearing and gets the properties of his/her father and possesses wealth. He/She will get name and recognition. He/She has a very competitive nature that makes a success. He/She is honest, calm and good at managing things and excelling at jobs with lot of travelling. His/Her marital life will be happy and contented. He/She has good sense of humour and good speakers. His/Her early life may be plagued by hardships restraints and possible impediments.

Male Natives

1. Physical features: He is tall, stout with short hands comparing to the body structure. There will be a scar mark on his upper right hand or beneath the shoulder.

2. Character: He has the magnetic power to attract others and gets respect and honour from the public. He helps the needy and does not deceive others. He is simple. His life is full of frequent ups and down due to some hidden curse on him.

3. Education, sources of earning/profession: He keeps strict discipline and works at managerial level. He/She is engaged in business or, in a high position in the service. He even with a preliminary academic background will possess excellent all-round knowledge. It will bring in unexpected circumstantial changes both in the family front as well as in his academic and professional or business field up to 30th years of age. He/She will settle down in his life between 30 to 42 years of his age. There will be remarkable accumulation of wealth and all-round success in the business beyond 64 years of age.

4. Family life: He is able to enjoy an ideal married life. His wife is a good housewife. His wife may indulge in homosexual activities in her youth period.

5. Health: He is prone to cough, cold, asthma or sinuous.

Female Natives

1. Physical features: She is extremely beautiful and entirely different body.

2. Character: She has shyness of a female sex. She respects the elders and will not hesitate to express her views openly, even; she is subjected to enmity from her relatives.

3. Education, sources of earning/profession: She is generally not employed or employed in the agricultural field and in the construction labour activities.

4. Family life: She can enjoy a happy married life. Her husband will be wealthy and loving. She enjoys good benefit from her children. There is one son and two daughters.

5. Health: Her health is good except high blood pressure and asthma in her old age.

Positive traits: He has ability to win people and situations over counts, exceptional sharpness, self control, faithfulness, and strength of intellect, innate ability to inspire and motivate people.

Negative traits: He is dominating and cruel sometimes and is prone to rude callousness and has lack of sensitivity.

Career: He shines exceptionally in careers related to entrepreneurship, counselling and consultancy and business, skill, craftsmanship and wisdom, and makes his way up the ladders of competition to reach top notch ranks and designation in technical lines and art. With his ability towards reconciliation he does better in mediating and settlement of disputes.

Ideal Partner by Nakshatra: This table gives you the ideal matchmaking and Compatibility between a Girl and a Boy for Marriage or between Partners for running a Business.

1st Partner Nakshatra	2nd Partner Nakshatra	Match / Synastry Status
Hasta (Kanya Rasi)	Bharani – Kritika – Mrigashira – Punarvasu – Ashlesha – Purva Phalguni - Uttara Phalguni – Chitra (1&2nd Pada) – Visakha(4th Pada) – Jyeshtha – Purva Ashadha – Uttara Ashadha (1st Pada) – Dhanishta (3&4th Pada) - P. Bhadrapada – Revati	Uttham
Hasta (Kanya Rasi)	Pushya - Magha - Anuradha - U. Bhadrapada	Madhyam

2.4.14　Chitra

General: He/She is well dressed, courageous, popular, expert in politics, rich, endowed with wife and children, God fearing and pious and defeats his enemies. He/She may pursue higher education. He/She is extremely knowledgeable and very concerned with political and social issues. He/She will do research, teaching and intellectual pursuits. He/She has very beautiful eyes with well-proportioned bodies. He/She is wonderful conversationalist and says the right thing at the right time. Many artists are born in this Asterism.

Male Natives

1. Physical features: He is lean and is identified even in a crowd of hundreds of people through his magnificent dealings and expressions.

2. Character: He is peace loving. He goes to any extreme for self benefits. He has great intuition and is very cordial with others. He confronts his enemies at every step but he is capable of escaping any conspiracy. He has soft corner for the downtrodden people and he devotes his time and energy for the uplift of this section of the society.

3. Education, sources of earnings / profession: He overtakes all hurdles and courage with hard work. He will not be leading a comfortable life up to the age of 32 years. 33 years to 54 years of age will be his golden period. He gets help and reward from unexpected quarters without putting many efforts. The age of 22, 27, 30, 36, 39, 43 and 48 years will be very bad in all respect for him. He may earn as a sculptor or mechanic or as a factory employee or a textile technologist.

4. Family life: He sincerely loves his co-born, and parents. He cannot enjoy the benefit, love and affection from his father. His father leads a separated life. He has a life away from his father. His father has distinct identity of his own and is well known. He cannot stay in the house where he is born. Either he will leave the house or the house where he is born will be sold or destroyed. In other words, he will be settling down at a distant place away from his native house. He cannot enjoy a happy married life. He has to shoulder lot of responsibilities and face lot of criticisms throughout his life.

5. Health: He may suffer from inflammation of kidney, bladder, brain fever and diseases connected with worms. Abdominal tumours or appendicitis also have been noticed.

Female Natives

1. Physical features: She has a beautiful tall body. She has natural long hair.

2. Character: She is proud, veracious, sinful and lazy. She commits sinful deeds.

3. Education, sources of earnings / profession: She will have her education in science subjects. She may be a nurse or a model or film extras. If planets are not well or moderately placed, she may be employed in agricultural field.

4. Family life: There is likelihood of death of the partner, divorce or complete absence of pleasure from her husband. In some cases, childlessness is also noticed.

Positive traits: He has an ability to create, build and appreciation of intrinsic beauty with an aura of polished glamour, refined and attractive mannerisms, power of intuition and ability to inhabit a mystical realm count amongst his positive traits.

Negative traits: He has intrinsic tendency to hide behind a veiling facade, selfishness, indulgence, dubiousness in his nature and does self service rather than generosity.

Career: He has career related to engineering, gardening, and horticulture, interior decor, glamour and hospitality industry.

Ideal Partner by Nakshatra: This table gives you the ideal matchmaking and Compatibility between a Girl and a Boy for Marriage or between Partners for running a Business.

1st Partner Nakshatra	2nd Partner Nakshatra	Match / Synastry Status
Chitra 1 & 2nd Pada (Kanya Rasi)	Aswini – Kritika – Rohini – Aridra – Pushya – Magha – Hasta – Anuradha – Moola – Shathabisha	Uttham
Chitra 1 & 2nd Pada (Kanya Rasi)	Revati - Visakha – Purva Phalguni – Ashlesha – Punarvasu – Bharani	Madhyam
Chitra 3 & 4th Pada (Tula Rasi)	Aswini – Kritika – Rohini – Aridra – Pushya - Hasta - Swati – Moola – Sharvana	Uttham
Chitra 3 & 4th Pada	Revati – Visakha – Purva Phalguni – Jyeshtha =	Madhyam

| (Tula Rasi) | Ashlesha – Punarvasu – Bharani | |

2.4.15 Swati

General: He has immense knowledge and excellent communication skills. He/She is long lived, rich, merciful, religious, and soft spoken and receives wealth from the king. He/She cannot be corrupted or manipulated. He/She may incur a lot of debt. He/She will do well in professions related to medicine and drugs, chemicals and travel industry. He/She tends to focus too much on social work and this may create friction within the family. He/She is rich, renowned and has prosperity, success. His childhood days will be full of problems.

Male Natives

1. Physical features: His under part of the feet is curved and the ankle risen. His feature is very attractive to the women folk. His body will be fleshy type.

2. Character: He is a peace loving but adamant. Once he loses temper, it will be very difficult to calm down. He is the best friend in need and worst enemy of the hated. He does not hesitate to take revenge on the persons who is against him.

3. Education, sources of earning/profession: He suffers, financially and mentally, even if born in a wealthy family, till his 25th years of age. He cannot progress in profession or business up to the age of 30 years. Thereafter he has a golden period up to his 60 years of age. He will earn through the profession as Gold Smith, travellers or drug seller, an actor or dramatist or may join defence (navy)and as astrologer or a mechanical engineer.

4. Family life: His married life is not much congenial and it is not adjustable couple.

5. Health: He has a very good health except piles and arthritis.

Female Natives

1. Physical features: She is very slow in walking.

2. Character: She is sympathetic and loving, religious, truthful, virtuous and enjoys a very high social position and has many friends. She will win over enemies.

3. Education, sources of earning/profession: She gets employment with much fame but does not like to travel. She is forced to accept more travelling due to job.

4. Family life: She will enjoy complete satisfaction from family and her children.

5. Health: Her internal constitution is weak. She may suffer from bronchitis, asthma, breast pain, broken feet ankles and uterus trouble.

Positive Traits: He is very knowledgeable, witty, very much independent, God fearing and religious minded, helpful, truthful in nature and competent; and has a high self respect and posses a strong liking to work in freedom.

Negative Traits: He is stubborn, adamant, and not ready to listen on work with some people. He has restlessness and uncertainty in decision making and is not very popular in the society.

Career: He is excellent and gets success in the later part of his life and not in the earlier stages. He has career interest in the financial and legal domains, gold business, acting, textile, travelling, mechanical engineering and astrology as well.

Ideal Partner by Nakshatra: This table gives you the ideal matchmaking and Compatibility between a Girl and a Boy for Marriage or between Partners for running a Business.

1st Partner Nakshatra	2nd Partner Nakshatra	Match / Synastry Status
Swati (Tula Rasi)	Bharani – Mrigashira (3&4th Pada) – Punarvasu – Ashlesha - Jyeshtha - Purva Ashadha – Purva Phalguni – Chitra - Visakha – Revati	Uttham
Swati (Tula Rasi)	U. Bhadrapada – Uttara Phalguni – Uttara Ashadha – Kritika - Pushya - Magha - Moola - P. Bhadrapada - Dhanishta (1&2nd Pada)	Madhyam

2.4.16 Visakha

General: He has power, position, and authority and leads life with his own principles and possesses his belief in Gandhian thoughts of non violence. He/She is communicator, writer, and speaker. He/She is very popular among the opposite sex. He/She has a strong sense of justice. His/Her careers will be in journalism, sales and marketing, writing. He/She has a very happy marital life. He/She is very goal oriented, and doesn't give up until achieve success. He/She is patient, persistent and determined. He/She experiences success in the second half of life.

1. Physical Features: He may be fatty and long structure or very lean and short structure.

2. Character: He does not believe in the orthodox principles or the age-old tradition and is fond of adopting modern ideas. Mostly he lives away from his family. Slavery is suicidal for him. He treats all religions, castes and creed as one and follows of Gandhian philosophy of 'Ahimsa. He accepts Sanyas (saints) when he touches 35 years of age and simultaneously looks after the family and follows Sanyas.

3. Education, sources of earning/profession: He is a very good orator and has the capacity to attract crowd. Hence he is the fittest person to be in the political circle. He does an independent business or job involving high responsibility, banking and religious professions or mathematician or a teacher.

4. Family life: He cannot enjoy love and affection of mother, may be either due to mother's death or unavoidable circumstances. He is always proud of his father. He leads more or less a life of an orphan. There is a lot of difference of opinion between his father and him and due to these reasons he is, right from the childhood, a hard working and self made person. He loves his wife and children very much. He is addicted to alcohol and indulges in too much sex with other ladies.

5. Health: His health is very good, but prone to paralytic attack after the age of 55 years.

Female Natives:

1. Physical features: She is extremely beautiful and faces lot of problems due to this.

2. Character: She has a very sweet tongue and expert in the household activities and in the official activities. She does not believe in pomp and show and is very simple.

3. Education, sources of earning/profession: When Moon and Venus are together, she may become a famous writer. She will have academic excellence in arts or literature.

4. Family life: She treats her husband as her god. Her religious attitude will confer love and affection from her in-laws.

She looks after the welfare of all family members and relatives. She may frequently be visiting sacred places.

5. Health: She has good health, but prone to kidney trouble, goitre and weakness.

Positive trait: He is very sharp and keen to learn new things, and do not like orthodox procedures and beliefs. He is full of energy and enthusiasm and quite single minded.

Negative traits: He gets involved with unsocial elements and practice like drug abuse, sex and alcohol habit. He is quite restless in nature and never feels contented with his achievements. This leads him to frustration and continuous worry.

Career: He is natural good orator and has a tendency to enter the political arena. Other careers are independent business, high responsibility jobs like those of administrative arena and teaching profession or mathematician.

Ideal Partner by Nakshatra: This table gives you the ideal matchmaking and Compatibility between a Girl and a Boy for Marriage or between Partners for running a Business.

1st Partner Nakshatra	2nd Partner Nakshatra	Match / Synastry Status
Visakha 1, 2, 3rd Pada (Tula Rasi)	Aswini – Mrigashira – Aridra – Pushya – Magha - Chitra – Swati – Moola – Dhanishta (1&2nd Pada).	Uttham
Visakha 1, 2, 3rd Pada (Tula Rasi)	Revati – Hasta– Purva Phalguni – Ashlesha – Rohini – Bharani - Anuradha - Jyeshtha – Dhanishta (3&4th Pada) - Shathabisha	Madhyam
Visakha 4th Pada (Vrishchika Rasi)	Aswini – Mrigashira – Aridra – Pushya – Magha - Chitra - Swati – Anuradha – Moola – Dhanishta – Shathabisha	Uttham
Visakha 4th Pada (Vrishchika Rasi)	Jyeshtha – Hasta – Purva Phalguni – Rohini – Bharani - Ashlesha - Revati	Madhyam

2.4.17 Anuradha

General: He/She dictates and has propensity in communications. He/She is destroyer of his/her enemies, famous, experts in arts, server of king, stationed in countries other than his/her own, truthful and respects his/her mother. He/She is fit for friendship, kind hearted, helping nature, enjoyments, traveler, fun loving, determined, spiritually inclined and enemy killer. He/She makes a lot of friends. He/She may go far away from place of birth because of profession. He/She may be successful in business. Careers that will suit are sales and marketing, astrology, and counselling. He/She will have a satisfactory marriage. He/She has superb leadership and organizational skills. He/She knows how to share and accommodate others. He/She has been known to live far from place of birth. There are many opportunities for travel. He/She has a peculiar frustrated face and a gloomy appearance because he has to confront on several occasions. There will not be peace of mind in his/her life. Even a smallest problem will start pinching his/her mind repeatedly. He/She always thinks of taking revenge whenever opportunity comes. In spite of these drawbacks he/she is the most hard working and ever ready to complete the task. After several reversals he/she ultimately achieves the desired result. He/She is a firm believer of god. His/Her life is full of helplessness but independent.

Male Natives

1. Physical features: He will have a beautiful face with bright eyes. In some cases where combination of planets is not good, he has cruel looks.

2. Character and general events: He is liable to face several obstacles in his life. Even then he has special aptitude to handle the most difficult situation in a systematic way.

3. Education, sources of earning/profession: He has a special calibre of how to pocket his superiors. He starts earning his bread at quite young age say on or about 17 or 18 years of age. He will have good life between the periods of 17 years to 48 years of age. His life after the age of 48 years will be extremely good. It is in this period he can settle down in his life to the desired way and become free from most of the miseries of the life. If Moon is in the company of Mars, he may be a person dealing with drugs and chemicals or a doctor.

4. Family life: No benefit will be derived from the co-born under any circumstances. His spouse will have all the qualities of a good housewife. He likes to provide all necessities of life as far as possible and love and affection to his children. Hence his children reach a high position, much more than that of him.

5. Health: His health will generally be good, but prone to asthmatic attack, dental problems, cough and cold, constipation and sore throat.

Female Natives

1. Physical features: She has attractive face, beautiful body and beautiful waist which attract a man.

2. Character: She is pure hearted. She likes to lead a simple life, a selfless, agreeable and attractive disposition. She can shine well in social and political field. She will respect her elders. Her friends will cordon her as if she is the head of friend's circle.

3. Education, sources of earning/profession: She is interested in music and fine arts. She may obtain an academic degree or high degree in music, dance or other professions.

4. Family life: She is very much devoted to her husband like Radha and observes religious norms. She can be called as a model mother as far as upbringings of her own children are concerned. Her devotion to her in-laws pours further glory in her personal life.

5. Health: She may suffer from irregular menses. Severe pain will be felt at the time of bleeding, as the flow of spoiled blood will be quite intermittent.

Positive traits: He/She has amiability, balance, good conduct, leadership ability, sensitivity to people and situation, and outstanding ability to forge rapport and harmony and is spontaneously efficient in bridging differences and animosity.

Negative traits: He/She has over emphasis on secrecy, frivolousness and unexpected mood swings, idleness, weakness and absence of purpose.

Career: His career opportunities are revolving around mathematics, science and engineering, tourism and travel and administrative job with an insistence on organizational skill.

Ideal Partner by Nakshatra: This table gives you the ideal matchmaking and Compatibility between a Girl and a Boy for Marriage or between Partners for running a Business.

| 1st | 2nd | | Match | / |

Partner Nakshatra	Partner Nakshatra	Synastry Status
Anuradha (Vrishchika Rasi)	Rohini – Punarvasu – Ashlesha – Hasta –Swati – Visakha - Shathabisha – Sharvana – P. Bhadrapada (1,2&3rd Pada)	Uttham
Anuradha (Vrishchika Rasi)	Revati - P. Bhadrapada - Jyeshtha – Chitra – Uttara Ashadha (2,3&4 Pada)- Uttara Phalguni – Magha – Mrigashira – Kritika - Aswini	Madhyam

2.4.18   Jyeshtha

General: He/She has material richness, achievement, kindness and hot tempered nature. He/She is endowed with few children. He/She achieves success early in life at the age of twenty-one. Careers that will suit are engineering, Architecture and construction. Married life is happy. He/She will have very supportive spouses. He/She is wise and the patriarch or matriarch of the family and knows how to deal with wealth and power. i) If born in the first pad, he/she is full of lustre and splendour, achieves fame and greatness, and is rich, brave, a hero and an excellent conversationalist; (ii) If born in second pad, he/she is of very cruel nature and quarrelling nature; (iii) If born in third pad, he/she follows the evil path and (v) If born in forth pad, he/she, although, has many sons but because of the influence of this malefic Nakshatra, it causes anguish and pain.

Male Natives

1. Physical features: He has a very good physical stamina, defective teeth bulging out.

2. Character: He is very clean and very sober. He cannot maintain secrecy or anything hidden in his mind even if it pertains to his own life. He is hot tempered and obstinate and forms a wall to his progress in life. He takes spot decisions without seeing the opportunity and circumstances, which ultimately leads him to a precarious state. He will not hesitate to cause problems and troubles to those who rendered him all possible help when required. He is prone to drugs and alcohol and goes out of control quickly damaging family life.

3. Education, sources of earning/profession: He will leave his home at a very young age and seek refuge in a distant place. He earns his bread with his own effort. Constant change of jobs or professions is noticed. Life will be full of trial up to 50 years of age and stability starts only after that. There will be maximum trouble from 18 years of age to 26 years of age and he will have to undergo financial problems, mental agony or mental disarrangement. There will be a beginning of progress towards stability from 27th years of age at very slow pace up to his 50 years of age.

4. Family life: He cannot expect any benefit from his mother and co-born because they become his enemies. His near and dear ones generally dislike him. He likes to keep a separate identity and existence. His spouse will always have an upper hand. There are health problems of his wife or separation due to some unavoidable circumstances.

5. Health: There is frequent temperature, dysentery, cough, cold, asthmatic attach and stomach problems. He may have severe pain in arms and shoulders.

Female Natives

1. Physical appearance: She has long arms, height above average, broad face, short and curly hair.

2. Character: She has strong emotions, passionate jealousies and deeper loves. She is intelligent, thoughtful, perceptive and good organiser.

3. Education, sources of income/profession: She is active in sports. She will get medium academic education. She is often contended herself in the home, seeking to enrich her own life through her husband's success. She not involved in any earning job.

4. Family life: She lacks marital harmony and notices loss of children. She is subject to harassment by in-laws. Her neighbours and relatives put poison in her life. She is always worried and her children will neglect her.

5. Health: Her health will not be so good. She suffers from disorder of the uterus. She is also prone to prostate gland enlargement or pain in arms and shoulders.

Positive traits: He/She is hard working, result oriented, willing to shoulder responsibility and position, and provides supportive protection to the weak and unsheltered persons.

Negative traits: He/She has temper and ego, hot headedness, obstinacy, and unwillingness to compromise.

Career: Career opportunities are military at managerial post, investigation, protection of law and order, and entrepreneurship.

Ideal Partner by Nakshatra: This table gives you the ideal matchmaking and Compatibility between a Girl and a Boy for Marriage or between Partners for running a Business.

1st Partner Nakshatra	2nd Partner Nakshatra	Match Status
Jyeshtha (Vrishchika Rasi)	Kritika – Mrigashira – Punarvasu – Pushya – Uttara Phalguni - Chitra – Visakha – Anuradha – Dhanishta	Uttham
Jyeshtha (Vrishchika Rasi)	U. Bhadrapada - P. Bhadrapada - Sharvana - Uttara Ashadha - Hasta – Swati - Purva Phalguni – Rohini – Bharani	Madhyam

2.4.19   Moola

General: He/She is close to politics & power, wealthy, famous, kind hearted and having long life. He/She is financially successful and leads a materially comfortable life. The work area is gardening, medicine, travel, tourist or agricultural. He/She gives good judgement. He/She enhances the prosperity and comforts his/her mothers. He/She is sage-like, wealthy and has vehicles, does permanent jobs and defeats his/her enemies. But, if born in the initial three parts of this Nakshatra, he/she causes destruction to his family. (i) If born in the first pad, he/she causes loss to his father, (ii) If born in the second pad, he/she causes loss to his mother and (iii) If born in the third pad, he/she causes loss to the family wealth.

Male Natives

1. Physical features: He has good physical appearance. He will have beautiful limbs and bright eyes. He will be the most attractive person in his family.

2. Character: He has a very sweet nature and is a peace loving person and has a set principle. He can stand and penetrate against any adverse tidal wave and reach the destination. He is not bothered about tomorrow. He keeps all the happenings in the hands of god. There will be frequent changes of profession or trade with no stability. He is always in need of money. He does not earn anything by illegal mode. He believes that all that is taking place on the earth is due to the blessings of God.

3. Education, sources of earning/profession: He has much better success in a foreign land and earns his livelihood in a foreign place in the field of fine arts or a writer.

4. Family life: He cannot have any benefit from his parents whereas he is all self-made. His married life will be satisfactory. His spouse has all the requisite qualities of good wife.

5. Health: He will be affected with is tuberculosis, esonophilia, and paralytic attack. There will be some severe health problems in his 27th, 31st, 44th, 48th, 56th and 60th years of age. He may be addicted to drug and intoxication.

Female Native:

1. Physical features: She will have reddish colour and her principal teeth will not be close, but at great distance, which is wealthy sign.

2. Character and general events: Mostly he is pure hearted, but very much adamant. Since she lacks knack of dealing, she quite often lands into problems.

3. Education, sources of earning/profession: She does not acquire much education. She doesn't show any interest in studies. She spends more than two terms in the same standard or class. Ultimately she leaves further education. The only exception is that if Jupiter is placed in opposition i.e. aspects or placed in Magha Nakshatra, she may be a doctor or employed in an envious position i.e. she will have excellent academic record and reach to the top.

4. Family life: She cannot enjoy full conjugal bliss, but a separated life, mainly due to the death of her husband or divorce. This result cannot be blindly applied as other planetary positions of favourable nature will nullify the bad effects. There may be delay in the marriage and also some hurdles. If the position of Mars is unfavourable she will have to face a lot of problems either from her husband or from children.

5. Health: She is prone to rheumatism, backache or pain in arms and shoulders.

Positive traits: He/She is hard working, committed, and intelligent to innovate ideas to make the road to success, highly optimistic and can come out of the toughest of situations.

Negative trait: He/She is provocative, stubborn and adamant. He/She has lack of knowledge to deal with things which often lands him into various problems.

Career: He/She does well in providing religious and financial advice, self business or self employment, in the foreign soil where he has a high possibility of success as compared to the native land. He/She switches his jobs frequently and possesses varied career interests.

Ideal Partner by Nakshatra: This table gives you the ideal matchmaking and Compatibility between a Girl and a Boy for Marriage or between Partners for running a Business.

1st Partner Nakshatra	2nd Partner Nakshatra	Match / Synastry Status
Moola (Dhanu Rasi)	Aridra – Pushya – Purva Phalguni – Hasta – Swati – Shathabisha	Uttham
Moola (Dhanu Rasi)	U. Bhadrapada – Visakha – Chitra – Uttara Phalguni – Punarvasu – Mrigashira (3&4th Pada) - Purva Ashadha - Sharvana - Dhanishta	Madhyam

2.4.20   Purva Ashadha

General: He/She is hugely popular in the society for his helping nature and possesses extraordinary argument capabilities. He/She has leadership abilities, strong sense of justice and has a beautiful wife. He/She has many friends and is an excellent manager. He/She has a forgiving nature and does not hold grudges against others. Careers that will suit are financial planning, social work, NGO's volunteer work. His/Her married life is satisfactory. He/She could suffer from ailments like Tuberculosis and uterine problems. He/She is red in complexion. He/She will travel overseas. He/She is wealthy, tall, good speakers, endowed with attractive face, obedient and lives with respect. He/She is loved by people, and experts in almost all fields. He/She is always successful and success comes at an early age. Great oratory abilities make him/her successful at debates.

1. Physical features: He is lean and tall. His teeth will be very beautiful, eyes bright, waist narrow and arms long. In other words, he has good attractive physical features.

2. Character: Nobody can defeat him in argument. He has extra-ordinary convincing power. He will not under any circumstance subdue to others, whether he is right or wrong. He can give a lot of advices to others. In decision-making, he is very poor, but too much dominant. He hates external show. He is God fearing, honest, humble and far from hypocrisy. He will be highly religious and devotes much of his time in Puja or others religious act. He is good collector of antiques. He may also take interest in writing poems.

3. Education, sources of earning / profession: Even though he can shine in almost all the fields, he is particularly fit for doctor's profession or fine arts. He is a not fit for any business. It will be period of trial and error up to 32 years of age. Thereafter he slowly starts climbing ladder of success. It is very good Period between 32 years to 50 years.

4. Family life: He cannot enjoy any benefits from his parents. However, he will be lucky to have benefits from his co-born, particularly from his brothers. He will be spending most of his life in foreign land. His married life is happy. His marriage may be delayed. He is more inclined towards his wife and in-laws. He has the most talented and respectful children, who will bring name and fame to his family. There may be at the most two children.

5. Health: His health is not good. He is prone to severe whooping cough, breathing trouble, bronchitis, tuberculosis, heart attack and malaria or esonophilia.

Female Natives

1. Physical features: She is extremely beautiful and magnetic. She will have long nose and graceful look, fair complexion, brown coloured hair.

2. Character and general events: She has energy, enthusiasm, vigour and vitality. She is greedy and is not obstinate. She will come to a final decision after deep consideration. She is straight forward. She will be fond of dogs and other pet animals. She makes promises but will not be fulfilled. She hates to her parents and brothers. She will be leader among her relatives. She is a determined, truthful character.

3. Education, sources of earnings / profession: She is educated. She may be a teacher, bank employee or attached to religious institutions. If mercury is placed with Moon, she may earn as a publisher or writer.

4. Family life: She is very good to her family and more attached to her husband. Benefit from the children will be to a limited extent only.

5. Health: Her health will be good. She will have acute disorder of the womb and uterus.

Positive trait: He/She possesses helpful and empathetic behaviour and is confident, self made and true to himself and the world. He/She provides with a lot of advice and ideas, which are helpful to people many a times.

Negative traits: He/She is arrogant, poor decision maker and cannot be deterred from doing what he has decided to do once.

Career: He/She has a variety of career interests, like medicine and fine arts. He/She is not recommended for ventures where important decision making is needed. Fields of science and philosophy are good career interests from the success point of view.

Ideal Partner by Nakshatra: This table gives you the ideal matchmaking and Compatibility between a Girl and a Boy for Marriage or between Partners for running a Business.

1st Partner Nakshatra	2nd Partner Nakshatra	Match Status
Purva Ashadha (Dhanu Rasi)	Mrigashira – Punarvasu (1,2&3rd Pada) - Magha – Uttara Phalguni – Chitra - Visakha – Jyeshtha – Moola - Uttara Ashadha 1st – P. Bhadrapada – Revati	Uttham
Purva Ashadha (Dhanu Rasi)	Aridra - Ashlesha - Punarvasu (4th Pada) - Hasta - Swati - Uttara Ashadha (2,3&4th Pada) - Sharvana - Dhanishta	Madhyam

2.4.21 Uttara Ashadha

General: He/She possesses good leadership skills and goes up to a respectable position in administrative services and has a beautiful and supportive life partner. He/She is very honest, plain hearted, very simple, straight forward and do not like hypocrisy and show off. He/She has political leadership ability. Careers that will suit are research, teaching, counselling, banking, notoriety and publishing. He/She will have a happy marriage and prove to be a devoted spouse. He/She may have more than one marriage. He/She is happy with his sons. He/She cannot lie is successful, and victorious. He/She does physical exercises, loved by people, and travels a lot. He/She is giver, gets enjoyment from wife and endowed with many children.

Male Natives

1. Physical features: He is a well proportioned body, broad head, tall figure, long nose, bright eyes, charming and graceful appearance with fair complexion.

2. Character: He gives respect to all and does god fear. There will be black mole around his waist or on the face. He will not deceive or cause any trouble to others. He will not take any hasty decision. Even in the state of conflict he cannot utter harsh words directly to any person. He i shoulders many responsibilities at a young age. He is subjected to maximum happiness at one stage and the maximum unhappiness at the next.

3. Education, sources of earning/profession: He has to be very careful in making any collaboration, otherwise, failure is certain. There will mark all round success and prosperity after the age of 38 years.

4. Family life: His childhood is better. His married is very good between the age of 28 and 31 years with a responsible and loving wife. The health of the spouse will be a cause of concern for him. His wife will be having problems of acidity or uterus disorder. He lacks happiness from the children, who will be main cause of concern.

5. Health: He is prone to stomach problems, paralysis of limbs, pulmonary diseases.

Female Natives

1. Physical features: Her forehead will be much wider, nose larger, eyes attractive, teeth beautiful, body stout but not so beautiful hair.

2. Character: She is the "obstinate daredevil". Her utterances are not good and jumps into conflict with others, but leads a very simple life. She is very well fit to the proverb "There are two horns for the rabbits I have got".

3. Education, sources of earning/profession: Generally, she is educated. She may be a teacher, bank employee or attached to religious institution. If Mercury is placed with Moon, she may earn as a publisher or a writer.

4. Family life: She does not enjoy married life due to either separation or some other problems connected with husband. She observes all the religious formalities. She derives complete happiness and contentment by marrying a Revati or Uttara-Bhadra Pad boy.

5. Health: She will have wind problems, hernia or uterus problems etc.

Positive traits: He/She is very much humble and respects others regardless of his social or financial positions. He gives utmost respect to women. He is very modest and doesn't utter anything bitter against anyone in spite of having strong opinion difference.

Negative traits: He/She lacks motivation to do things in life and expect appreciation of his work, failing which he feels unhappy and depressed.

Career: He/She achieves success in career after 38 years of age. He may have to face initially some struggle. The fruitful career interest includes engineer, architects, mechanical engineering jobs and working with maps and planning.

Ideal Partner by Nakshatra: This table gives you the ideal matchmaking and Compatibility between a Girl and a Boy for Marriage or between Partners for running a Business.

1st Partner Nakshatra	2nd Partner Nakshatra	Match / Synastry Status
Uttara Ashadha 1st Pada (Dhanu Rasi)	Aridra – Pushya – Magha - Purva Phalguni – Hasta – Swati – Anuradha – Moola - Purva Ashadha – Shathabisha – U.	Uttham

	Bhadrapada	
Uttara Ashadha 1st Pada (Dhanu Rasi)	Aswini - Bharani - Mrigashira - Ashlesha - Jyeshtha - Sharvana - Dhanishta - Revati	Madhyam
Uttara Ashadha 2, 3, 4th Pada (Makar Rasi)	Aswini – Bharani – Pushya – Magha – Purva Phalguni – Hasta – Swati - Anuradha – Moola - Purva Ashadha – Sharvana – Shathabisha – U. Bhadrapada	Uttham
Uttara Ashadha 2, 3, 4th Pada (Makar Rasi)	Rohini - Ashlesha - Jyeshtha - Dhanishta - Revati	Madhyam

2.4.22 Sravana

General: He/She is truthful, endowed with many children and friends, winning over his enemies, large-hearted, knowledgeable, learned, wealthy and famous, and pure in heart. He/She is loved by all. Sravana bestows with immense wealth of knowledge, a lot of respect to their parents and caring of them with the utmost sincerity. He/She is experts in various art forms like music, dance, and drama, and acting. They do not ever hurt or pose problem to anyone wilfully. They are extremely religious, pious. He/She is good speaker, enjoys life and has many sons, organizational leadership skills. He/She is financially successful, and leads a materially comfortable life. He/She will be a good manager or project leader and will do well in business. He/She works for charitable organizations. Careers that will suit are engineering, management related jobs, charitable and social institutions. Marital life will be extremely happy and blissful. He/She is egoistic by nature. He/She can get service in education field and may be great teacher, or perpetual students. Counselling is a gift and he/she has the ability to truly listen.

Male Natives

1. Physical features: He has a very good attractive physical feature. His height is small. His face may have a mole or some other marks appearing to be a kind of disfigurement. He has black mole beneath his shoulder.

2. Character: He is very sweet in the speech and principled and expects his surroundings to be very clean and neat. He takes pity on the condition of others and tries to help them as far as possible. He is a very good host. He is God fearing and has full Guru Bhakti. He is a believer of 'Satyameva Jayate' (Truth only will win). Neither he will reach to the top nor will he be at the bottom. He will shine well. Since he has to shoulder many responsibilities and spend for fulfilling his responsibilities he will always be in need of money.

3. Education, sources of earnings / profession: He will undergo several changes up to 30 years of age. He will mark stability in all walks of life between 30 years to 45 years of age. He can expect remarkable progress both economical and social beyond 65 years. He will take up mechanical or technical work or engineering connected with petroleum.

4. Family life: His married life will be filled with extra ordinary happiness. He will have most obedient wife with all good qualities. He has sex relationship with other ladies.

5. Health: He suffers from ear problem, skin disease, eczema, rheumatism & tuberculosis.

Female Natives

1. Physical features: She is tall and lean. Her head is comparatively big with broad face. She has large teeth. There will be distance between the front teeth with prominent nose.

2. Character and general events: She believes in charity. She is highly religious and visits several holy places. She is internally very cunning. However, sympathy towards the weak and generosity towards the needy are the notable features. She is a 'chatter-box' without having any control over her tongue.

3. Education, sources of earning / profession: She is Illiterate or medium educated and will be engaged in agricultural field, or employed as typists, clerks or receptionist etc.

4. Family life: She is a pride in family but has frequent friction with her husband.

5. Health: She has skin disease and prone to eczema, filarial, pus formation, and tuberculosis, leprosy of low intensity and ear problem.

Positive Traits: He/She is large hearted and compassionate, and helps people with his knowledge, which makes him lovable among people. He/She has courage, forgiving gratitude and a comprehensive personality in the society and respect.

Negative Traits: He/She is shrewd, selfish and adamant and has wrong steps or unorganized efforts and harms others for the sake of accomplishing their target.

Career: His career spans through services to having his own business and is successful in career. He/She is interested in engineering, medicine, education, science and many other forms of arts. He/She is financially sound unless he goes to the wrong path.

Ideal Partner by Nakshatra: This table gives you the ideal matchmaking and Compatibility between a Girl and a Boy for Marriage or between Partners for running a Business.

1st Partner Nakshatra	2nd Partner Nakshatra	Match / Synastry Status
Sharvana (Makar Rasi)	Bharani - Mrigasira – Punarvasu – Ashlesha – Uttara Phalguni (2,3&4th Pada) - Chitra – Purva Phalguni - Vishakha – Jyeshtha Purva Ashadha - Uttara Ashadha – Dhanishta – P. Bhadrapada - Revati	Uttham
Sharvana (Makar Rasi)	Magha - Purva Phalguni - Uttara Phalguni (1st Pada) - Anuradha - Moola - U. Bhadrapada	Madhyam

2.4.23 Dhanishta

General: He/She has symphony, prosperity and adaptability, ability in music dance, confidence, stability, dependability, hard work, energy, exceptional sharpness, commercial skills and benevolence, fame, success and prosperity in abundance, striking a comfort zone with the given surrounding, luxury and good life and group centric activities like an artist. He/She is difficult to win, very famous, serves the elders and always protect others. He/She is religious, endowed with many good qualities, wealthy, kind hearted, sad and suffers from diseases like tuberculosis. He/She is after other sex as is loved by them. He/She is a glib talker, skilled in business and astrology. Marital life will be happy and satisfactory. He/She is dark in complexion, cheerful, have all materialistic pleasures, soft-spoken and well mannered. The work area is Ayurveda, mining and engineering in underground works, land, and business of any product, commission agent, metal related work and machinery. He/She is visionaries and good for public relations. Marriage may be delayed or denied. He/She gains fame, recognition and loves travel.

Male Native

1. Physical features: He has a lean body with lengthy figure and stout figure.

2. Character: He has extremely intelligent mind and all-round knowledge. He does not cause any trouble to others and has religious spirit. He is very revenge taking and, if any person causes trouble to him, he waits for an opportunity to bounce upon him.

3. Education, sources of earning/profession: He is the born scientists and historians. He has an inherent talent of keeping secret; he is quite suitable for secret service, private secretaries to the senior executives. His intelligence is beyond questionable. Hence lawyer's profession is the best for him. It will show progress in the earning field from 24th years of age onwards. He should be very careful before putting trust on others.

4. Family life: He will be the uppermost administrator. His relatives will cause a lot of embarrassment and problems. He is more inclined to his brothers and sisters. He will have ample inherited property subject to the placement of planets in beneficial position. He cannot have much benefit from his in-laws. His wife will be an incarnation of 'Lakshmi' (goddess of wealth). There will be improvement in his finance only after the marriage.

5. Health: His health is not so good. He is prone to whooping cough, anaemia.

Female Natives

1. Physical features: She is beautiful and ever sweet seventeen while she crosses her forty or may have ugly appearance due to teeth protruding outside the lips.

2. Character: She has sympathy towards the weak. She is an enforcement master like Sharvana. She has congenial atmosphere in the home front.

3. Education, sources of earning/profession: She has mixed talents and education. Hence usually, she will be teachers or lecturers or research fellows.

4. Family life: She will be an expert in the household administration.

5. Health: Her health is not good. She is prone to anaemia, uterus disorders of acute intensity and spoiled blood

Positive traits: He/She has vibrant mannerisms, frankness and easy adaptability, and striking a bond of geniality and harmony with the immediate surroundings, hopefulness, joy and sympathy.

Negative traits: He/She is susceptible to the society and may manifest subsequent negativity in his behavioural traits and has aggression, talkativeness, materialistic ways, covetousness, lust for success and susceptibility to select incompatible life partners.

Career: Careers are related to performing arts, exceptional niche in managerial positions and catering to group activities. Thus career opportunities based on management and entrepreneurship are also suitable for them. They are found to be equally prosperous in medical profession; particularly in specialized branch of surgery, military bands, real estate and scientific research.

Ideal Partner by Nakshatra: This table gives you the ideal matchmaking and Compatibility between a Girl and a Boy for Marriage or between Partners for running a Business.

1st Partner Nakshatra	2nd Partner Nakshatra	Match Status
Dhanishta 1 & 2nd Pada (Makar Rasi)	Aswini – Krittika – Pushya – Uttara Phalguni (2,3&4th Pada) - Hasta – Swati – Anuradha - Moola – Uttara Ashadha - Sharvana – Shathabisha	Uttham
Dhanishta 1 & 2nd Pada (Makar Rasi)	U. Bhadrapada – Purva Ashadha – Visakha – Ashlesha – Punarvasu -	Madhyam

	Krittika (2,3&4 Pada) - Jyeshtha - Uttara Phalguni - Magha	
Dhanishta 3 & 4th Pada (Kumbh Rasi)	Krittika - Pushya - Magha - Uttara Phalguni - Hasta - Swati - Anuradha - Moola - Uttara Ashadha - Sharvana - Shathabisha	Uttham
Dhanishta 3 & 4th Pada (Kumbh Rasi)	Purva Ashadha - Jyeshtha - Visakha - Purva Phalguni - Ashlesha - Punarvasu (4th Pada) - Aridra - Rohini – Aswini, U. Bhadrapada	Madhyam

2.4.24   Shathabisha

General: He/She has care, remedy and healing, intellect and power of intuition, reclusive loneliness, scientific attitude, meditation and reflection, depression and mood swings, strict adherence to discipline and rigid compliance with norms and seldom going beyond the norms. He/She lives in foreign lands and is most sexy. He/She is respected by the world, long lived, well in transactions and suffers from various diseases. He/She is endowed with many children, most selfish, gambler, questionable character, experts in magic, stubborn, especially knowledgeable, and cruel, eat little, wealthy, server, and destroys his enemies. He/She has a noble and an aristocratic demeanour, and can become dangerous when provoked. Careers that will suit are Engineering, Accounts, and medicine and surgery. Marital life is not without friction. He/She is dark in complexion, weak body and has saving habits. He/She may be healers or doctors. He/She can be astronomers as well as astrologers. He/She may be alcoholic and may suffer from diseases difficult to heal or hard to cure

Male Natives

1. Physical features: He will have excellent memory power, wide forehead, attractive eyes, bright countenance, prominent nose and bulged abdomen. He appears to belong to an aristocratic family at the first sight itself.

2. Character and general events: He is of the type 'Satyameva Jayate'. He will not hesitate to sacrifice his own life for upholding the truth. He is born with certain principles; he has to quite often confront with others as he cannot deviate from his principles of life. Selfless service is his motto. He does not believe in pomp and show and has interesting and attractive conversation, which will be highly instructive and educative.

3. Education, sources of income/profession: It will be a trial period up to 34 years of age and after 34 years, it will be the period of constant progress. He practices astrology, psychology and healing arts. His literary capacity and greatness will come to limelight even when he is very young. He is capable of acquiring very fine and high education. He will be eminent doctors and research fellows in medicines.

4. Family life: He faces a lot of problems from his dear and near ones. He always helps to his near and dears. He undergoes to maximum mental agony due to his brothers. He cannot also enjoy much benefit from his father, whereas full love and affection is derived from his mother. He does not have a happy married life. In some cases, when there is severe affliction of Saturn and Jupiter, it has been found that he remains a chronic bachelor throughout his life, but his wife will have all the good qualities of a companion.

5. Health: He is prone to urinary diseases, breathing trouble and diabetics. He is too much inclined in the sexual pleasures and may have sexually transmitted diseases. He may keep illicit relationship with other females. He may have problems with his jaws. He is also liable to suffer from colic troubles.

Female Natives

1. Physical features: She is tall, lean, fairly beautiful, elegant disposition, fleshy lips and broad cheeks with prominent temples and buttocks.

2. Character: She is religious and god fearing. She has hot-temperament and mostly confronts with family quarrel and lacks mental peace. She has very good memory. She is highly sympathetic and generous.

3. Education, sources of income/profession: She has scientific study as a doctor.

4. Family life: She loves her husband, but life is full of problems due to long separation from her husband or widowhood.

5. Health: Her health is not good. She suffers from colic, chest pain, Urinary and uterus.

Positive Traits: He/She has hard work, determination, discipline, ability to cure and heal; powers of intuition and meditation, educational excellence, methodical approaches together with pleasing skills of presentation and powerful memory.

Negative Traits: He/She is susceptible to obstinacy, adamant and head strong, and have farfetched seriousness and flair for loneliness, bouts of angry outbursts and depression, unfriendly mannerisms, inadaptability, long drawn insistence on tradition and inability to innovate and has laziness, rudeness and absence of social skills.

Career: He/She shines exceptionally in careers, on account of his academically oriented nature, memory and skilled faculty of intuition, revolving around medicine, psychology, astrology and astronomy and due to owing to the predominating influence of Rahu, natives are prone to making careers out of politics, business and in those areas where leadership skills are required.

Ideal Partner by Nakshatra: This table gives you the ideal matchmaking and Compatibility between a Girl and a Boy for Marriage or between Partners for running a Business.

1st Partner Nakshatra	2nd Partner Nakshatra	Match Status
Shathabisha (Kumbh Rasi)	Mrigasira – Punarvasu – Ashlesha – Purva Phalguni – Chitra – Visakha – Jyeshtha – Purva Ashadha - Dhanishta -	Uttham
Shathabisha (Kumbh Rasi)	Revati – P. Bhadrapada – Uttara Ashadha – Moola – Anuradha – Uttara Phalguni - Pushya - Punarvasu - Aswini	Madhyam

2.4.25   Purva Bhadrapada

General: He/She has mystery, supernaturalism and occult phenomena, honesty, principle and benevolence, sincerity, determination, discipline and ability to confront all kinds of physical and mental hardships. He/She is worshiper, teachers, broad minded, respectable, above controversies, soft spoken, and loving his/her relatives. He/She is endowed with children, sleeps excess, helping nature, and has the knack of trade. He/She, most likely, will be employed in a government organization and will gain good promotions. He/She has quite independent life both socially and financially. It will mark remarkable all-round progress in his life between 24 and 33 years of age. It will be his/her golden period between 40 years and 54 years of age, when he/she can establish fully. He/She keeps much restriction on the spending activities. He/She can shine in the field of business, banking, government job, or as a teacher, actor or writer, research worker and astrologer or astronomer.

Male Natives

1. Physical features: He has a lifted ankle of the foot, medium size with fleshy lips.

2. Character and general events: He is peaceful, loving, very simpleton type, very much principled, and likes good food and is a voracious eater. He normally expresses impartial opinion. He does not believe in the blind principles of religion. He is ever ready to lend a helping hand to the needy. Even so, hatred and resistance will be his reward in return. He is financially weak. He is God fearing and performs religious rites. He is moderately rich, but likes to have respect and honour from the public rather than accumulating money.

3. Education, sources of earnings / profession: He has the knack of trade and shines in any type of job he undertakes. He is employed in a government organization with unexpected gains or promotions in the revenue collection department or where cash transactions take place.

4. Family life: He cannot enjoy full love and affection from mother. Mother is away most of the time and he is automatically separated. He can be proud of his father. His father has fame in the field of fine arts, oration or in the writing field and possess a very good moral character. In spite of these good qualities of his father, he quite often confronts with him.

5. Health: He is prone to paralytic attack, acidity and diabetes. He may have problems with the ribs, flanks and soles of the feet.

Female Natives

1. Physical features: She is very beautiful in appearance and has slim figure.

2. Character: Honesty and sincerity are her main characteristics. She believes that let her head is cut down, but she will not deviate from the right path and principles. She is a born leader. She is capable of extracting work from others. She will be successful when she gets power and authority. She has the humanitarian doctrine, and politeness.

3. Education, sources of earnings / profession: Her education will be in the scientific or technical field and becomes teacher, statistician, and astrologer or research worker.

4. Family life: She is more attachment towards her husband and will be blessed with children. She has good ability in the house hold administration. She will enjoy lots of benefits from her children. She will have many children if she marries a Rohini boy.

5. Health: She is prone to low blood pressure, dropsy or swollen ankles, apoplexy and palpitation, perspiring feet and enlarged liver.

Positive traits: He/She has sincerity, dutifulness, endurance and strength, helpfulness, protectiveness and concern for acquaintances and family.

Negative traits: He/She has innate tendencies to cause harm and malice. He/She is prone to socially destructive ways and terrorism. Being misled into dishonesty, negativity and immorality also count amongst his negative traits.

Career: His careers relate to administrative service, business, and governmental responsibilities and teaching profession; in scientific research, writing and acting. Lines related to astrology, astronomy and analysis of scriptures are also beneficial for him.

Ideal Partner by Nakshatra: This table gives you the ideal matchmaking and Compatibility between a Girl and a Boy for Marriage or between Partners for running a Business.

1st Partner Nakshatra	2nd Partner Nakshatra	Match Status
Purva Bhadrapada 1, 2, 3rd Pada (Kumbh Rasi)	Aswini - Mrigasira (1&2 Pada)– Pushya – Magha- Chitra – Swati – Anuradha - Moola – Dhanishta - Shathabisha	Uttham
Purva	U. Bhadrapada – Sharvana –	Madhyam

Bhadrapada 1, 2, 3rd Pada (Kumbh Rasi)	Purva Ashadha – Jyeshtha – Anuradha – Hasta - Ashlesha	
Purva Bhadrapada 4th Pada (Meena Rasi)	Mrigashira – Aridra – Chitra (1&2 Pada) – Anuradha - Moola – Dhanishta – Shathabisha – U. Bhadrapada	Uttham
Purva Bhadrapada 4th Pada (Meena Rasi)	Sharvana – Purva Ashadha – Jyeshtha - Hasta – Pushya - Swati	Madhyam

2.4.26 Uttara Bhadrapada

General: He/She is wise, prominent amongst his/her clan, wears jewels, a lawyer or arbitrator, always does constructive jobs, earns wealth, giver, endowed with children, religious, winners of his/her enemies, happy, determined, sexy, learned, great sacrificed, loved & respected in all circles, fickle, and suffers from fluctuation of funds. He/She is rich, famous, attractive, charming, loyal to friends, follows the virtuous path, always willing to come to aid and will always stand out in a crowd. He/She learns to face the world with confidence. He/She has the 'possession of lucky feet', and represents prosperity, strength and martial qualities, material bondage and rainfall. Malefic or the benefic aspects of ruling Saturn plays an important role in determining his personality. He/She has determination, wisdom and experience, changeability, attracting virtue and mannerisms and highly ethical standard. He/She is intelligent financially successful and self-sufficient. He/She is a loving and merciful person and always willing to reach out to others and spend a lot of time and money on charity. He/She is good at solving problems. Marital life will be happy, harmonious and satisfactory and children will be a source of joy and happiness. There is wealth usually as a gift or inheritance, usually latter in life. He/She is extremely protective of his/her loved ones. His/Her wisdom seems to have magical powers. He/She has happy home life and is blessed with good children.

Male Natives

1. Physical features: He is most attractive and innocent looking person. There is an inherent magnetically force in his look. If he looks at a person with a mild smile, rest assure, that person will be his slave.

2. Character: He keeps equal relationship with high and low people status of the person. He has a spot-less heart. He does not like to give troubles to others. He has temper always on the tip of his nose. However, such short-temper is not of a permanent nature. He will not hesitate to sacrifice even his life to those who love him. At the same time once he is hurt he will become a lion. He has wisdom, knowledge, and personality. He is expert in delivering attractive speeches. He is capable of vanquishing his enemies and attains fairly high position. He is sexually inclined always and desirous of being in the company of other sex.

3. Education, sources of earning/profession: He can attain mastery over several subjects at the same time. He is academically much educated and his expression and knowledge put forward to the world will equal to that of highly educated and learned persons. He is much interested in fine arts and has ability to write prolonged articles or books. Laziness is a remote question for him. Once he opts to undertake a job he cannot turn back till that job is completed. He is employed and reaches to the top. Even born in poor family, he is employed and reaches to a good position and he always receive reward and praise from others. His stability in life or the slightest upward movement begins after his marriage. He starts his livelihood at a very young age say 18 or 19 years of age. He will have important changes in the professional field at his 19th, 21st, 28th, 30th, 35th and 42nd years.

4. Family life: He cannot virtually derive any benefit from his father. He leads a neglected childhood. He is normally subjected to a life away from his home town. His married life will be full of happiness. He will be blessed to have a most suitable wife. His children also will be an asset, most obedient, understanding and respecting. He will be blessed with grandchildren also. He is an ornament in his family.

5. Health: His health will be very good. He is non-care about his own health. Hence he will search for a doctor only when he is seriously ill. He is prone to paralytic attack, stomach problems, piles, and hernia.

Female Natives

1. Physical features: She is of medium height with stout body. She will have large and protruding eyes.

2. Character: She is a real 'Lakshmi' (goddess of wealth) in the family. Her behaviour is extremely cordial, respectful and praise worthy and has adaptability.

3. Education, sources of earning/profession: Employed females can attain good positions due to her own effort. She is best suited to the profession of a lawyer or arbitrator. She is also a good nurse or a doctor.

4. Family life: She will be a gem in any family, where she is born or married. In other words, their footsteps are sufficient to bring in Lakshmi (goddess of wealth).

5. Health: She is prone to rheumatic pains, acute indigestion, constipation, hernia and in some cases tuberculosis of low intensity.

Positive traits: He/She is kind, logical, and deft at skills of calculations, and has serenity and sense of equality and justice. He/She has intrinsic innocence along with an innate capacity for mesmerizing people and a unique potential for gaining with experience.

Negative traits: He/She has susceptibility to fits of anger and is unnecessarily drawn into controversy and disputes and has temperamental, laziness, inertia and loss of control are their other negative traits.

Career: With wealth of knowledge gained by experience, he makes favourable impression on business, entrepreneurship and managerial positions. Career opportunities centring on consultancy, justice and counselling are also meant for him.

Ideal Partner by Nakshatra: This table gives you the ideal matchmaking and Compatibility between a Girl and a Boy for Marriage or between Partners for running a Business.

1st Partner Nakshatra	2nd Partner Nakshatra	Match Status
Uttara Bhadrapada (Meena Rasi)	Rohini – Aridra – Punarvasu (2&3 Pada) – Hasta – Jyeshtha - Sharvana - Shathabisha - P. Bhadrapada – Revati	Uttham
Uttara	Dhanishta – Uttara Ashadha	Madhyam

Bhadrapada (Meena Rasi)	- Moola - Swati – Ashlesha - Uttara Phalguni (3&4th Pada)- Punarvasu (4th Pada) – Krittika (2,3&4th Pada)	

2.4.27 Revati

General: He/She as power to transcend, idealism, benevolence and sustenance, helpfulness, elements of graceful sociability; streaks of knowledge and educational excellence. He/She reaches to people and has sympathetic approach to people and situations which gives commanding awe, reverence and love. He/She is endowed with good qualities, always thinking about home, wise, wealthy, enjoys life, learned, fickle, earn good money, sexy, givers, pure minded, capable, bright, helps others and moves in foreign lands. He/She is extremely likeable and easy to get along with. He/She will enjoy great prosperity. He/She has amicable nature, controls his senses, acquires wealth and possesses sharp intelligence. He/She enjoys being a robust physique and good health. He/She should be careful with finances or could incur a lot of debt. His/Her careers will suit in Counselling, Psychology, Psychiatry, Social work and volunteer work in spare time. Marital life will be very harmonious and spouse is very compatible. He/She is fair in complexion, peaceful, Pandit, good working, soft spoken, famous in family, wealthy, and enjoys family values. He/She gets very angry. He/She has an affinity or love towards small animals. He/She may have disappointment early in life. He/She is a bit weak and may be prone to childhood illnesses. He/She loves water, and usually benefits from living by the water. There is a deep devotion and faith to God.

Male Features

1. Physical features: He will have very good physique, moderately tall and symmetrical bodies with fair complexion.

2. Character: He is very clean in his heart, sincere in his dealings and soft-spoken. He cannot keep anything secret for too long time. He will not blindly believe even his loved ones. But once he takes somebody into confidence, it is not easy to keep away them out of his attachment. He is very hot tempered. He tries to observe the principle, which he feels, is correct and resist tooth and nail till the end. He is God fearing, superstitious, religious and rigid in the observance of orthodox culture and principles. Hence he also enjoys the maximum blessings of the Almighty.

3. Education, sources of earnings / profession: He studies historical research and ancient cultures like astrology and astronomy. He is a good physician or poet. He is employed in a government organization with the most success. He is mostly settled in foreign countries at quite a reasonable distance from one's own birthplace. He will come up in his life with his own efforts. His intelligence and abilities are the inherent qualities at birth. He cannot stick to any particular field of job for long time. He cannot gain much till his 50th years of age. It will be a good period between 23 years and 26 years whereas period between 26 to 42 years of age will mark a lot of problems financially, mentally and socially. It is only after his 50th year he can think of worry less and stable life.

4. Family life: He cannot get help from his relatives or father. He is unlucky to enjoy the benefits from his near and dear ones. However, his married life will be moderately good. His spouse will be quite an adjusting type.

5. Health: He is prone to fever, dysentery or dental diseases, and intestinal ulcers.

Female Natives

1. Physical features: She will be extremely beautiful. She can be recognized easily even out of thousand ladies due to her magnificent attractive personality.

2. Character and general events: She is somewhat stubborn. She likes to exercise authority over others. She is God fearing, religious and rigid in the observance of orthodox culture and principles and highly superstitious.

3. Education, sources of earnings / profession: She will have her education in the field of arts, literature or mathematics. She will be in the general line i.e., she may be a telephone operator, typist, teacher or a representative of companies. When good aspect of benefice planet is received, she may be an ambassador or a person representing her country for cultural or political matters.

4. Family life: She will enjoy a most harmonious married life.

5. Health: She may have some deformities of the feet, intestinal ulcers or abdominal disorders. In some cases, deafness has also been noticed.

Positive traits: He is pillar of strength and support to people and society at large, oriented towards religion, intrinsically constructive by nature and has ethics and principles, appreciable sense of decorum, sophistication and culture, and optimism. Ability to reach out and serve others being the major fulcrum of his behavioural characteristics, he is often seen holding magical sway over others.

Negative traits: He is susceptible to idleness and depression and has strict adherence to principle and unexpected norms of behaviour obstinacy, temper, head strong ways, orthodoxy and superstition.

Career: Career involves scientific research, archaeological survey; poetry, literature in ascendance. He can also be successful in careers related to administration, astrology and astronomy due to his wisdom and social acumen.

Ideal Partner by Nakshatra: This table gives you the ideal matchmaking and Compatibility between a Girl and a Boy for Marriage or between Partners for running a Business.

1st Partner Nakshatra	2nd Partner Nakshatra	Match / Synastry Status
Revati (Meena Rasi)	Krittika (2,3&4th) – Mrigashira – Punarvasu (1,2 &3rd Pada) - Uttara Phalguni (2,3&4th Pada) Chitra (1&2nd Pada) - Visakha – Anuradha – Uttara Ashadha – U. Bhadrapada	Uttham
Revati (Meena Rasi)	Shathabisha – Sharvana – Visakha – Hasta – Pushya – Purva Ashadha - Punarvasu (4th Pada) - Rohini – Krittika (1st Pada)	Madhyam

3

House (Bhava)

3.1 General

The Sanskrit name of the House is "Bhava" or "Sthana". The Houses govern all the "events" of our lives. If we look directly east and point our hand at the Eastern horizon where the Sun rises, we are pointing to the first House, which is called Lagna or Ascendant. Each House/Ascendant/Lagna occupies the 30-degree span of space below the Eastern horizon. Directly across from it or 30-degree span of space above the Western horizon is the 7th house of the Natal Chart. From the first to the 7th are the 2nd through 6th houses. Directly over our head is the 10th house, which is called MC. That is the straight up into space where the high noon Sun beams down on us. Straight below our feet is the 4th house. Suppose 9 O' clock position of the clock represent the Eastern horizon. So, the first house governs at 9.00 to 8.00 O' clock position of the clock. The 2nd house is just below it, i.e. at 8.00 to 7.00 O' clock position of the clock and so on. Thus, the 12th house governs at 10.00 to 9.00 O' clock position of the clock, and this brings us back around to the first house of the Natal Chart. Thus, the entire 360 degree circular span of space surrounding us are divided into 12 equal part of 30 degree sections each.

3.2 Concept of House and Lord

It is like that we are standing inside a circular room. The circular room is divided into eleven rings up to the wall, starting from the ring nearest to the wall to the ring at the centre of the room. There are nine Planets, one in each ring. The native is in the central ring, which represents the Earth. The ring nearest to the wall has 12 divisions, each representing one Sign. The native is standing at a point in the central ring, i.e. on the Earth which is fixed. The other nine rings containing one Planet each and the tenth ring containing 12 Signs are moving around the native. The Planets are moving at different speeds. The wall of the room is fixed and is divided into 12 parts, each representing one House. When native looks around he sees the 9 Planets and 12 signs rotating within the room. Whenever he looks at any one of the planets, he sees a particular coloured portion of the Sign and a House behind them from his viewpoint. Thus, each planet in a certain Sign is in a certain House always. It does not mean that the planet is actually in those Sign at that time, but simply that the particular Sign is the backdrop far away behind the planet.

In Vedic astrology, one sign always occupy one house. Whatever sign is raising the entire portion of sign is considered in the first house. In this way there is an exact correlation between Sign, Sign Lord and House. The Lord of the House is the Lord of the Sign in that House. The lord of the first House and its position in the other house will together determine the body. The body gets shape of the house in which the First House ruler has come to stay in and that house greatly affects the body. The matter related to the body is affected during periods ruled by the lord of 1st House in the Vimshotari Dasa. Secondly, the planets associated with the 1st lord greatly affect the affairs or events of body of the native. Example: If the lord of the first House is placed in the first house; the body becomes the focus of the life. If the lord of the first house is placed in the second house then the body will be greatly influenced by the second house, i.e. possession and wealth. The native is busy with accumulating wealth whether he is coming from a strong family background or otherwise. If the lord of the first House is placed in the third house, then the body is found to be connected with younger brothers, sisters, and the intelligence, good education, college degrees and other things that the third house rules. If the lord of the first house is placed in fourth house then the person will be greatly concerned with family affairs, home, parents and house & land properties in his life. The native will spend a lot of time at home, rather than on the road or at work. Since the fourth house is tenth from the seventh house, it also means that the native will have some connection with the career of the spouse. This means that the person might work for spouse or work with her spouse or perhaps a home-based business. If the lord of the first house is placed in the fifth house then the children, creativity, romance and helping nature becomes the focus of his bodily activities. If the lord of the first house is placed in the sixth house, then the deaths, diseases and enemies overwhelm the person. If the lord of the first house is placed in the seventh house, then the person is heavily focused on his partner, his spouse during his life. If the lord of the first house goes to the eighth house in a horoscope, then serious physical harm; unforeseen difficulties and serious problems comes to the native. If the lord of the first is placed in the ninth house then there is an overall fortunate protective cover on the native throughout his lifetime. If the lord of the first is placed in the tenth house, then the native is heavily focused upon the attainment of career, status position and success. If the lord of the first is placed in the eleventh house, then the person is heavily focused upon the achievement of desires and certain gainful things in life. If the lord of the first is placed in the twelfth house, since this house rules charity, donations, and losses, the person will be busy in these ways throughout his life. In other words, he may donate himself to these causes. Or, there may be a lot of loss in his life.

Thus, the placements of the lords have their first level meanings. Then the second, third, fourth and deeper and deeper meanings depend on the astrologer's ability to read the complexity of the house relationships with the planets and signs.

3.3 Bhava (Chalit) Chart

House is important aspect in astrology. The Lagna Chart house limits are only approximately. Actual house position is governed only by Bhava Chart or Chalit Chart. Hence Chalit Chart cannot be ignored in astrology. Chalit Chart is used for predictions of effects of actual position of planet in house. However, the Lagna Chart house is used for determining the aspects, or knowing the sign in which a planet is posited or knowing the strength of a planet or any other purposes. For example, if a Sun position moves from 1st house of Lagna Chart to 12th house in Chalit Chart, while it was in Aries in Lagna chart, then it does not mean that Sun has changed the Sign and shifted to Pisces. It remains in Aries only. But it has changed the house cusp and is considered to be placed, actually, in 12th house for predictions of effects of planet positioned in 12th house. It is very important to have calculation of house of a planet by making the house division for correct placement of planets for correct predictions house wise. Because, the planets show their behavior according to the house actually they occupy.

The Lagna chart only tells us in which sign a planet is posited. It also tells us about the aspects of a planet on the other planets. However, it does not confirm the house in which a planet is situated in real sense. So we make another chart called Chalit, which gives us the information about the house in which a planet is posited. Normally, the planet position in the house of Chalit Chart and the house occupied in the Lagna chart are same. But, sometimes, they tend to differ when the ascendant degree is high and planet degree is low or vice versa. To represent this situation, we mark the houses with degrees and place the planets and houses in the order of the degree they own. If the planet degree lies in the extent of the house degree, then place the planet in that house, otherwise if a planet is beyond the limit of the house, it is considered in next house. It is very important to consider the house division for correct predictions, because the planets show their behavior according to the house, actually, they occupy. The behavior is enhanced depending upon its strength, which is determined depending upon the sign or Nakshatra it occupies. Hence use Lagna chart only to know the sign in which a planet is posited to know its strength and to know the aspects. Use Chalit chart to confirm position of the planet in a specific house for correct prediction.

3.4 Special Lagna in Vedic Astrology

Sage Parasara mentioned a few special Lagnas. He wanted special Lagna to be used for clear predictions of the events. Special Lagna is used for some clear purposes. As Lagna stands for self, Special Lagna stands for various *shades* of "self". Special Lagna is very important and is one of the very particular and magnificent methods of Indian/Vedic astrology. Each special Lagna shows a particular life area. Special Lagna age as below:

Arudha Lagna: Arudha Lagna stands for "manifestation of self". It deals with 'perceptions' and from the point of material world. It is a maya (illusion). It shows the "public image".

Chandra Lagna (Moon Sign): It is the Hose, where the Moon is sitting in the Natal Chart. It is also called Moon-Sign or Rashi in Hindu astrology and is very important Lagna for predictions. Chandra Lagna is the first house counted from the sign of the natal Moon, and is used to analyze mind, memory and mental activity, and also used to determine how fertile a woman will be. Moon is said to be the Lamp post of Destiny, the day to day predictions are given based on the position of moon. There is a great importance of Chandra Kundali/Moon chart in astrology. Moon is the fastest planet in all the 9 planets, after ascendant (Janma Lagna) it is the moon which moves faster, one ascendant (Janma Lagna) changes roughly in 2 hours 15 minutes, moon changes a sign in 2 and a quarter days roughly. Moon is karaka of mind; transits are to be seen from Moon and Moon chart. One should always refer to Chandra Kundali while reading Lagna Kundali. Chandra Kundali is an important tool of prediction and the results of planetary combinations are more prominent by Chandra Kundali. Should a malefic be in the 4th, identical with an inimical Rashi, counted from Chandra, while there is no benefic in a Kendra, the child will lose its mother in a premature manner. Among all the Lagnas, the Moon's Lagna (Chandra Lagna or the sign in which the Moon is posited), is most important for assessing the effects of transits (Gochara Phala). It is, therefore, imperative to make predictions about effects of the transits of planets through various signs from the sign occupied by the Moon. Chandra lagna forms the baseline for most emotional experiences in life, primarily kinship and marital relationships, specifically with parents and children.

Ghati Lagna Ghati Sign): It changes sign every Ghatee (every 24 min). Ghati Lagna (by calculation is for name and fame). Ghati Lagna stands for "self, from the point of view of power, fame and authority". If Ghatee Lagna is in a Dusthana (in 6, or 8, or 12) from natal Lagna, problems with your authority and social influence can be expected. Your authority will not be respected by other people easily, you'll have to struggle a lot, your power and influence is decreased. It is taken for predictions of name and fame. Ghati Lagna changes along with every Ghati (24 minutes) from the sunrise. Find out the birth time in Ghati and Vighati and consider the number of Ghati past, as number of signs, or Ghati Lagna. The Vighati be divided by 2 to arrive at degrees and minutes of arc, past in the said Ghati Lagna. The product so arrived in Rashi, degrees and minutes is added to Surya longitude, as at sunrise, to get the exact location of Ghati Lagna.

Hora Lagna: It changes sign every 2.5 Ghatee (each hour). Hora Lagna (by calculation is for financial prosperity). Hora Lagna stands for "self, from the point of view of financial prosperity". If Hora Lagna is in a Trikona (1 or 5 or 9) from natal Lagna, it gives better results for wealth and prosperity. Hora chart reveals for you that how wealthy you will be. The Hora Lagna Analysis is designed to analyze your financial future. In Vedic astrology, one of the important ascendants is Hora Lagna. When it comes to financial matters the placement of Hora Lagna is essential as it controls your wealth and your financial life. People are generally concerned with finances. It is taken for predictions of finance & prosperity. Hora Lagna remains in between day and night. From sunrise till the time of birth, Hora Lagna repeats itself every 2½ Ghati (i. e. 60 minutes). Divide the time from sunrise time to up to birth time by 2½ and add the quotient in Rashi, degrees and so on to the longitude of Surya, as at the sunrise. This will yield Hora Lagna in Rashi and degrees.

Indu Lagna: Indu lagna is a tool suggested by classics on astrology for estimating the wealth of the native. Here wealth is not simply currency. Prosperity in all aspects of life can be studied from this lagna. If a benefic planet aspects the Indu Lagna, it gives a lot of wealth in its Dasha. Many benefic planets aspecting the Indu lagna, wealth would be in abundance. If one or more planets are in their own signs or exaltation signs, would make the native even richer. Malefic exalted and occupying the Indu lagna makes one very wealthy at the end of their Dasa. If the malefic occupying the Lagna is not in its exaltation sign or own it also gives wealth but moderately. Planets occupying Trines and quadrants from Indu lagna give wealth in their Dasa/Antar Dasa. Debilitated malefic aspecting and or occupying the Indu lagna are not good for wealth. Jupiter in its transits comes into a trine or quadrant from this lagna also gives wealth. Planets situated in Dusthana (3, 6, 8, and 12) from this lagna make one incur loss of wealth. If the Indu lagna is not occupied or aspect by any planet, it is not good for acquiring wealth.

Indu Lagna or Dhan Lagna (Wealth Ascendant): Indu Lagna (by calculation is for wealth). Indu Lagna also known as Dhan Lagna or Wealth Ascendant is a special ascendant, which is derived through calculating cumulative position of ninth house from Ascendant/Lagna and the Moon. Calculation of Indu Lagna or Dhan Lagna (Wealth Ascendant): Some values are assigned to the planets, which are to be used in this calculation and are mentioned below:

Planet	Sun	Moon	Mars	Mercury	Jupiter	Venus	Saturn
Value	30	16	6	8	10	12	1

The planets Rahu and Ketu are not used in this calculation. Find out the zodiac sign (Rashi) falling in the ninth house of the Lagna Chart; and write down the value of the lord of ninth house. Then, find out the zodiac sign (Rashi) falling in the ninth house of the Moon Chart; and write down the value of ninth house lord in Moon Chart. Add both these values; and then, divide it by 12 and note down the remainder, which will be 0 to 11. Count the number from the position of the Moon; and the house identified will be the Indu Lagna or Dhan Lagna or Wealth Ascendant. If the remainder is zero, the house/sign falling in the twelfth house from the Moon or the previous house from the natal Moon's position will be considered as Indu Lagna or Dhan Lagna or Wealth Ascendant. If the remainder is one, the house/sign occupied by the Moon will be considered as Indu Lagna or Dhan Lagna or Wealth Ascendant. And, similarly the position of Indu Lagna or Dhan Lagna can be identified for the horoscope. After ascertaining the Indu Lagna or Dhan Lagna; the next step is to analyze the Indu Lagna Chart. As the name suggests, Dhan Lagna or Wealth Ascendant is meant for analyzing the financial prospects of a horoscope. The planets placed in the first, second, fourth, seventh, tenth and eleventh houses from Indu Lagna or Dhan Lagna are told to be the giver of wealth and financial gains to the native. The planets having their aspect on the first house of Indu Lagna or Dhan Lagna are also considered to be supportive for generating financial gains for the native. Debilitated, weak or malefic planets, in the above situation, are considered to be the destroyer of wealth. However, if such malefic planet is exalted or otherwise strong and/or receiving auspicious planetary influences; it can also generate financial gains. Planets disposed in third, sixth, eighth and twelfth houses from the Indu Lagna or Dhan Lagna are considered to be inauspicious for wealth and financial prospects of the horoscope. The results of these planets are experienced during their Dasa and/or Antardasa and in accordance with their strength and disposition.

Janma Lagna (Lagna Sign): In Natal chart, the first House is called Janma Lagna. There are Badhaka Graha defined as per the Janma-Lagna Sign. Even though the Badhaka Graha is natural Benefic, he works as the Badhaka Graha and harms badly that house in which he is sitting during his Maha Dasa and Antar Dasa.

Karaka Lagna (Karaka Sign): It is the Lagna/Ascendant taken from the Lordship of "Karaka Graha" and taken for predictions of the events related to that House with respect to the position of that Karaka Graha.

Karakamsha Lagna: It shows the most active and creative (create, kriya, kri, karak) behaviours in the present lifetime. In modern era, the Karakamsha usually shows highest expressions of profession.

Lagna (Ascendant): It changes sign every 5 Ghatee (every two hours). Therefore on every two hours we have another ascendant (Lagna). Lagna stands for "physically *true* self". The Ascendant or Lagna shows the self of the individual. The ascendant, "Lagna" is exactly the point in zodiac (signs), which rises at eastern direction at time of birth (or event or request or a "Prasana"). This rising is based on the daily rotation of the earth on its own axis, which determines "day" and "night". The earth needs 24 hours for this rotation. Lagna is therefore no star, no planet, but it is a very important 'symbolic point' in horoscope. It is of utmost importance in natal chart (Rashi Chart). On average, every two hours another sign (Rashi) is rising, therefore on average every two hours we have another ascendant (Lagna). The whole sign, containing this rising point, forms the 1st Bhava. Lagna in natal chart contains a summary of all life themes and areas, the basic and core personality, the body, the incarnated body of the eternal Soul and the self. Lagna indicates which life conditions an incarnated Soul will meet in this new incarnation, which surroundings with regard to Dharma, Artha, Kama and Moksha matters. Lagna, Tanu Bhava (Kendra, Dharma): It represents Physical stature, colour, form and shape, constitution, health, vitality and vigour, natural dispositions and tendencies, personality and struggle for life, honour, dignity, prosperity, general well being, head, upper part of the face, virtues, longevity, start in life and an idea about the general structure of life. The Rashi containing the Lagna Lord is called Paka Rashi.

Shri Lagna (Shri Sign): Sri Lagna (by calculation is for prosperity and marriage). Sri Lagna shows overall prosperity. In Sanskrit, the word "Shree" means wealth. It also means Lakshmi, wife of Narayana and goddess of wealth. Shree Lagna may be denoted by SL. Shree Lagna is important for prosperity. Shree Lagna in Western astrology is known as "Part of Fortune", a particular degree in Zodiac or Rashi which indicates luck when planets are Yuti or have Drishti with it. It is Lagna taken for predictions of prosperity and marriage. Sree Lagna is a place where Goddess Laxami (of wealth) resides. Sree lagna is considered to be one of the most important lagnas or special ascendants in Vedic astrology. Like all other special lagna, sree lagna also has the power to influence certain aspects of human life, considering its position in the natal chart of a native. However, the area that sree lagna is most predominantly linked is money or fortune matters. The sree lagna position in your birth chart determines the aspects of material gain and your attitude to the material possessions of life. For instance, a Vedic astrology horoscope reading for you, in respect of the sree lagna position as well as sree lagna influence, will explain to you how you regard the prospects of material gain. You also come to know about a possibility of gaining a fortune. Therefore, if you have a chance to make money, the awareness of it will help you take all necessary steps and make all possible efforts so that you do not miss the opportunity.

Surya Lagna (Sun Sign): Surya Lagna (ascendant counted from sign where the natal Sun is located). It is the House, where the Sun is supposed to be sitting in the Natal Chart. It is defined with the date of Birth of the native. It is called Sun-Sign in Hindu astrology and is very important Lagna for predictions. Sun sign at the time of the birth is known as Surya Rashi. Sun sign is the name of the zodiac in which the Sun was located at the time of the birth. Do you know your rising zodiac sign or in which Ascendant you are born? And, are you born in Leo Ascendant and wondering what your horoscope says about you? If your answer is, "Yes", this article may give you some good hints about the basic traits of your personality as well as some good understanding about the role and nature of different planets for your horoscope. Sun is a hot, masculine, dominant and kingly/ royal planet. It is not really a generic malefic unlike Rahu, Ketu, Saturn and Mars. Instead Sun can be and is life giving and warm in its nature. However, it is still considered a mild general malefic by nature due to its hot disposition and intense nature. When close to any planet, it makes the other planet invisible due to Sun's own intense brightness and this phenomenon in astrology is known as combustion. The combust planet loses some of its ability to do good in certain ways and loses Istha Phala, which has been explained already in another article on this site. Likewise, whichever bhav (house) Sun sits in, gets hot/ burnt to some extent at the least and hence this nature of Sun makes it a mild malefic in the context of Vedic Astrology.

Upa-pada lagna: Arudha of bhava-12. The A-12 profiles our most private reality, especially matters of the marital bedroom.

Varnada Lagna (Varnada Sign): It is Lagna taken for predictions of social company. We find that the Varnada Lagna is based on the Lagna and Hora Lagna. The Lagna indicates the native's health, appearance, actions, intelligence and skills while the Hora Lagna shows wealth, sustenance, longevity and his value system as what he considers valuable. We may infer from the use of these that the Varnada Lagna (VL) will provide additional information about the native's health, appearance, longevity, wealth and most important his career and social standing. It shows the native intelligence working on the value system and that karma which becomes expedient upon him to perform. Varnada placed in a Kendra to the Lagna will show a person who has the support of his society and family in initiating his career. If Varnada is occupied and owned by natural benefic planets then the native is strongly supported by the community and such family members indicated by the planets. If malefic planets conjoin or lord the Varnada, then strife, challenges and battles are indicated. If Moon is placed in Varnada Lagna and aspects the 9th house or lord, then career starts easily and very easily. If such a Moon is afflicted then there is considerable mental turmoil and work suffers.

Varnada Lagna placed in trines to the Lagna shows a person with good adaptability to any community, understanding and upholding the laws of that community. The Lord of Varnada, if a natural benefic and so placed shows that the native easily adapts into his career and is a strong support for his professional colleagues, but if afflicted he is tormented by his colleagues and life in the office can be very difficult. Varnada

Lagna in a Dusthana (sixth, eighth or twelfth house) from Lagna makes the native shy away from the community, enjoying a life in seclusion and avoiding regular social functions and the community. The reasons for the distancing are known from the nature of the lord of the Dusthana. Varnada Lagna in the twelfth house makes the native highly antisocial. This placement can, if other factors are present, show a demonical attitude towards his community, which would enable him to do unlawful acts, such as crime etc. If the lord is debilitated then the native is troubled by demonical people (showing a good soul) and if associated with Mars, will have to battle them in order to survive. If the dispositor is exalted or the lord has Nichabhanga Rajayoga then he will surely succeed in finishing the demonical opposition. In any case, if the Varnada is in the twelfth house, the native will be forced by circumstances to leave his home and fend for his fortune in another place. Varnada in the second house makes a person an honest worker and gives a high and admirable social standing which is conducive to Rajayoga. If the Lord is also a natural benefic and is associated with another benefic planet then Rajayoga is enjoyed due to family. The nature of the Lord of the Varnada shows the source of the Rajayoga. If afflicted then the native abdicates the throne. Varnada in 3rd house makes a native free of enemies and his relationship with the community constantly improving making the native happy and content. The native has parakrama (enterprise and daring) if associated with malefic planets. Similarly, if the Varnada is in the eleventh house, the community will be source of achieving the desires of the individual.

Vighatee Lagna: It changes sign every Vighatee (every 24 sec)

3.5 House-to-House Relationship

Houses are related to each others. The second house to any of the other house will indicate the wealth or money acquired from that house. The eighth house is second to the seventh house and will indicate the wealth and money of the spouse. The sixth house is the 2nd to the fifth house and will indicate the financial prosperity of the children. The eleventh house is the second house to the tenth house and will indicate financial gain through the career. Similarly, the twelfth house to any house is the end or loss to the house in question.

Example: The sixth house is the twelfth to the seventh (marriage), so it represents the end of the marriage or wife. The third house is the end of the mother because it is the twelfth to the fourth (mother), so it represents the end of the mother. The third house is our energy, will, and life force and the second house is the twelfth house from our life force and so is the loss of our life force. Thus, Maraka house (2nd) derives its meaning from this principle. It is the twelfth from the third. Similarly the eighth house is our length of life so the twelfth from the eighth, i.e. the seventh house would be the loss of life, i.e. the death house. So, 2nd and 7th houses are called Maraka houses.

The 4th house indicates about "mother". 5th house will indicate the mother's money as it is the 2nd house from 4th house. 6th house will indicate the mother's younger siblings as is the 3rd house from 4th and so on. 1st house will indicate the mother's Career, because this is the 10th house from mother's house. These are called the "compound" or "secondary" houses. The 12th house rules loss, and therefore, the 12th house from any house is the loss of that house. Take the 9th house which rules fortune, religion or Dharma, the father, the spiritual master and guru, and God's grace. The 9th house is 12th from the 10th house. This means that the house of Dharma or religion is the house of loss to the house of career, profession and position. The 10th house rules not only career, but mainly it rules rise and status in material life. It rules standing up tall and straight and getting some position, some fame, and some power in this material life.

Similarly, the 5th house is 12th to the 6th house. Amongst other things, the 5th house rules winning at the lottery and the 6th house rules debts, therefore, it is easy to understand that if we win at the lottery, we can cure all our debt problems. So, the 5th house, which is a money house, puts an end to the 6th house, the house of debts. The 6th house is 12th to the 7th house. The 7th house rules marriage, the spouse and partners in our life. Therefore the 6th house rules the loss of the partner or loss of the spouse.

The 7th house is 12th to the 8th house. The 8th house rules the vital source of energy, which is longevity and so, the 7th house rules the end of vital energy. The 8th house is 12th to the 9th house, which rules father. The 8th house, therefore, represents loss of father. Therefore, if we find the 9th lord in the 8th house in a chart, it is often found that the person lost his father.

3.6 Classification of House/Bhava:

Houses Categories	Houses	Effects of houses
Trikona (Trine) or Kona or Auspicious houses	1, 5, 9	They are the most auspicious or Benefic houses. They give fortune, luck, bring spirituality and well being if, unaffiliated.
Upachaya (Pratipas/Trika)	3,6,11	Upachaya means "improvement" and are considered auspicious. Life improves and gets better over time with these houses if, unaffiliated.
Dustasthana (Trikas) or Malefic houses	6, 8, 12	Trikas are the most in-auspicious or malefic and deal with suffering, ill health, disease, death, loss and sorrow. The rulers of these houses will inflict this type of suffering.
Bhoga	2, 4, 10	Bhoga are considered auspicious and deal with the pleasure and luxury in all respect such as Cars, furnished sweet home and other part of life.
Kendra	1, 4, 7, 10	Kendra is considered auspicious. The planets in Angles give effects in one's boyhood.
Panapharas	2, 5, 8, 11	The effects of planets in Panapharas are felt in the middle age.
Apoklima (Cadent)	3,6,9,12	The planets in Apoklima give result at the conclusion of the life, i.e. old age.
Maraka	2, 7	Maraka means "killer". The rulers are considered the killer sometimes. They are prominent when death or injury occurs.
Lakshmi	4, 10	The planets in Lakshmi gives home, happy life, conveyances,

		happiness, treasure, lands and buildings, heritage, real estate, good profession or livelihood or honour, living in foreign lands, reputation, business and social activities.
Vishnu	5, 9	The planets in Vishnu provide pleasures through children, love affairs and romance, knowledge, royalty or authority, and fun, long distance travel and make fortunate, and gain of spiritual knowledge,

Aims of Life	Houses Number	Effects of houses
Dharma	1,5,9	They relate to our sense of purpose and the spirit that moves us such as self through 1st; creative expression through 5th; and our spiritual beliefs and truths through 9th house.
Artha	2,6,10	They relate to our achievements, such as money and material through 2nd; hard work through 6th; and the public recognition and career through 10th house respectively.
Kama	3,7,11	They relate to our sense of conveying our ideas, needs, and desires, such as, the need of a life through 3rd; partnership through 7th; and feel connected to every one through 11th.
Moksha	4,8,12	They relate to our liberation or freedom of soul through 4th; past essence of the soul through 8th; and about releasing all attachments to the world through 12th.

Weak House: The House/Bhava gets weak under the following circumstances:

- When it gets sandwiched between malefic, particularly the natural malefic.

- If Bhava is occupied or aspect by the 6th, 8th or 12th lords (lord of Dustasthana).
- If the lord of the Bhava is in debility or combust.
- If the lord of the Bhava is influenced by the lords of the 6th, 8th or 12th house.
- If the lords of the Bhava occupy any of the 6th, 8th or 12th houses (Dustasthana) and is also aspect by the natural or functional malefic.

3.7 House Signification (Karaktva):

The Karaka Sign and the most Karaka Planet of the twelve Houses and their Signification are as given below. If the Karaka planet is in the house by position or aspect, that house gives very good effect to the native.

House	Karaka Sign	Most Karaka Planet	Modern title of house	Interpretation
1st	Aries	Sun	(Thanu/Tanu Bhava)-House of Self	Mind, Physique, Physical appearance, General outlook into the world.
2nd	Taurus	Jupiter	(Dhan/Vikrama Bhava) House of Money	Money, Self-Worth, Belongings, Property, Acquisitions.
3rd	Gemini	Mars	(Sahaj/Vikrama Bhava) House of Communications	Higher education, Communication. Siblings. Neighbourhood matters.
4th	Cancer	Moon, Mercury	(Bandhu Bhava) House of Property and Family Matters	Property, pleasure, Happiness, Conveyance, Enjoyments.

5th	Leo	Jupiter	(Santana/Putra Bhava) House of Children	Children, Knowledge, Creation, Enjoyment, Games, Gambling, Love and sex.
6th	Virgo	Mars, Saturn	(Ari Bhava) House of Health	Jobs and Employments, Health and Overall well-being, Service performed for others.
7th	Libra	Venus	(Yuvati/kalatra/ jaya Bhava) House of Spouse & Partnerships	Close relationships, Marriage and Business partners, Agreements.
8th	Scorpio	Saturn	(Randhra/Ayust han Bhava) House of Reincarnation	Deaths, Rebirth, Sexual relationships, Inhrited properties, Finances, Self-transformation.
9th	Sagittari us	Sun, Jupiter	(Dharma Bhava) House of Religion & Philosophy	Foreign travel, Foreign countries, Long distance travels and journeys, Religion, Law, Higher education, Knowledge.
10th	Capricor n	Mercur y, Jupiter , Saturn	(Karma Bhava) House of Profession & Social Status	Business, Ambitions, Motivations, Career, Status in society, Government,

				Authority, Father, Breadwinner of the household.
11th	Aquarius	Jupiter	(Labh Bhava) House of Gain & Friendships	Friends, Groups, Clubs and Societies. Higher associations, Benefits and fortunes from career.
12th	Pisces	Saturn	(Vyaya Bhava) House of Expense & Self-Undoing	Places of seclusion such as hospitals, prisons and institutions, including self-imposed imprisonments.

1st House (Thanu/Tanu Bhava): The First house or Lagna called Tanu Bhava represents body looks, soul, head/Brain, personality traits, longevity, health, character and nature, life style, complexion, inherent disposition, vitality, temperament, ego, paternal grandmother's wealth, maternal grandfathers' wealth, residence abroad, livelihood and pride. Benefic in the Ascendant will endow with servants, happiness and robes, while malefic give adverse effects in regard to these. It is one of the most important and auspicious house in the horoscope.

2nd House (Dhan Bhava): The Second house called Dhana Bhava represents wealth, money, quality of speech, family, face, right eye, mouth, food, charity, death, primary education, one's deposits, income, friends, sanyas, and security. It predicts material and financial resources and the ability to earn money, income, and inflow of finance, wealth possessions, precious stones possessions, domestic life, bank position, and understanding with family members, law suits and domestic comforts in general. The second house also projects any accident, nature of accident and how death will come. A benefic in the 2nd will, however, cause his/her death in during its mahadasha.

3rd House: Sahaja / Bhatru / Krama / Vikrama / Parakrama / Sahottha / Yodha Bhava): The Third house represents younger brothers and sisters, sibling, co-born, sports, throat and singing, voice, music, business, servants and subordinates, communications, higher secondary level education, talents and skills, short distance travels, neighbour, surroundings, relatives and relations with them; boldness, arts such as theatre or filmy arts, filmy direction, painting, drawing, and success through own efforts, competition, hearing and father in law. The third house rules menial work, cough, respiratory system and partition of property.

4th House (Bandhu Bhava): The Fourth house called Matru Bhava represents mother, home, relatives, office or factory, emotions, domestic and house related happiness and luxury, landed and house property, mental peace, chest and lungs, higher education such as master degrees, home affairs & home pleasure, possession of vehicles/conveyances, conditions at the old age, matters of the heart, inheritance and false allegations, pleasure trips, savings, cattle and pets.

5th House (Santana/Putra Bhava): The Fifth house called Putra Bhava represents children, intelligence and knowledge and intellect, creativity, mantra, tantra and pooja, speculation, love affairs, recreation, romance, creativity and stomach. 5th house represents teaching, principal and gynecologists, sports, relationships, luck, legal or illegal amusements, authority, and pregnancy. It predicts abortions, politics, good karma, destiny, lotteries, and self-projection in order to please prominent people or the public or a boss at a job. The fifth house indicates wanderings, arbitration, and higher education. It covers number of children, their longevity and their character, status, sex of children and education of children. It covers intuition, previous karmas and father's side of influence.

6th House (Ari Bhava): The Sixth house called Shatru Bhava represents health, illness, injuries, loans, sports, enemies and opposition, digestive system, quarrels, court cases, litigation, maternal uncle and aunt, servants, work environment, jobs & service, step-mother, imprisonment, medical profession, food, restaurants, subordinates, obstacles in life, mental worries, calamity, employee, hard work, fear from thieves or enemies, fighting, misery and success over enemies, loss of moneys, cheating, danger and calamities (troubles) through women.

7th House (Yuvati/Kalatra/ Jaya Bhava): The Seventh house called Kalatra or Juvati Bhava represents spouse, partners, sex, marriage, business, trade, employment in a private firm, sexual enjoyments, Kundalini Shakti, sexual organs and diseases thereof, death, relationships and signifies kidneys. 7th house represents married life, travel, conjugal happiness, loved one, divorce, honour, residence and reputation in foreign country, interaction with others, attitudes, sexual passions, open enemies, impotency, desire, disputes, relationships with wife, family life, age of getting married, wife age, health and her nature, journeys to distant places, loss through females, relationship and freedom.

8th House (Randhra/Ayusthan Bhava): The Eight house called Ayusthan Bhava represents longevity, destruction, accidents, physical pains, inheritance, legacies, death and reasons of death, underground wealth, historical things, monuments, parental property, failure, family of spouse, life's secrets, joint resources, anus and sex power. 8th house represents financial windfalls, lottery winnings and physical pains. It represents worries, finances through unfair means, obstacles, gain from in-law, imprisonment, struggles, Mafia & underworld, bankruptcy, obstacles, surgery, research, intuition, and long-term sickness, monetary gains from partner, misfortunes, trouble from enemies and loss of property. Malefic in the 8th will cause loss of spouse in the Dasa periods of the Lord of the Navamsa occupied by the 8th Lord.

9th House (Dharma Bhava): The Ninth house called Bhagya Bhava represents luck, prosperity, guru, father, religious and spiritual progress and knowledge of the scriptures, sadhana, pilgrimages, long journeys like foreign travel and foreign trade and dealing with foreigners, grandchildren, higher studies like doctorate, knowledge of foreign languages, grandparents, and signifies hips. 9th house represents fortunes, prosperity, writing books, powers of foresight, religious institutions, teaching, lawyers, fame, happiness, and unexpected gain of wealth, gain from lottery and affluence, association with good people, inclination toward God, religious and social work and fame in it, and satisfaction and fulfilment of desires.

10th House (Karma Bhava): The Tenth house called 'Karma' represents profession, business, authority, power, honours and achievements, acts (karma) one does, father, government service, politics, management, career, status, mother-in-law, prestige, reputation and signifies knees. It increases public image and makes the native's parent of greater influence, and provides power or authority like bosses, judges, and big stars. 10th house signifies popularity, status, activities outside house, pleasures, government favour, command, adopted son, worldly activities and moral responsibilities, livelihood, living in foreign lands, debts, reputation, social activities, social position, fame, wealth of the father, earned money, meritorious deeds, hardship in work, service, foreign place of settled life, promotion and number of promotion.

11th House (Labh Bhava): The Eleventh house called Labh Bhava represents gains, sources of income, elder brothers and sisters, friends, long distance travels, air plane travel, entertainment, friend's circle, daughter/son-in-law, accumulated wealth, ambition, social life, association and club, emotional attachments, son's wife, quadrupeds and attitude towards them, desire in life and colleagues or co-workers, love affairs and girl friends, honour and social success, clothes, the staple food, gold, and gain of wealth, pleasure, followers, dependent, insight, power of overcoming obstacles, redemption, worth of garments and signifies legs.

12th House (Vyaya Bhava): The Twelfth house called Vyaya Bhava represents losses, waste, expenses, long journeys like foreign countries and residence in foreign countries, imprisonment, death, sadhana and Moksha or final liberation, bad habits, hospital, quests, export and import, feet, sleep, donation, foreign stay, subconscious, psychological issues, secrets disputes, and signifies mental agony, bodily injury. 12th house covers one's own death, good food & comforts, bed and couch pleasure, donations, miseries, sufferings, troubles, betrayals, law suits, imprisonments, hospitalisation, conjugal relations with opposite sex other than wife, contacts, misfortunes, secret enemies, spiritual liberation, sea or ocean travel, interest in arts & films, renunciation and enjoyment. The twelfth house also indicates the foreign trips, number of the foreign trips, benefits or loss due to the foreign trips and settled in the foreign lands, divine favour and travels.

4

Zodiac Sign

4.1 General

The Zodiac is a band of group of stars or the positions of celestial bodies. The Zodiac is divided into twelve divisions of 30 degrees each called "Sign".   Each segment is called a Sign (Rashi).

Gender (Male/Female) Sign: The alternate Sign starting from Aries onwards are known as male and female on the other hand. The Gender of the Sign will help the Astrologer in assessing one's children, brothers and sisters in terms of Males and Females, from the horoscope.

Directions of Signs: The four Signs from Aries onwards indicate East, South, West and North, while the remaining Signs repeat in the same way. A journey undertaken by a person towards the direction indicated by the Lagna or the Moon at the commencement of journey yields fruitful results.

Night and Day Signs: Gemini, Cancer, Capricorn Aries, Taurus and Sagittarius are night Signs. Leo, Libra, Scorpio, Aquarius, Pisces and Virgo are day Signs.

Strength of Signs: If a Sign is aspect by its Lord, or by a planet friendly to its Lord, or by Mercury, or by Jupiter, it is said to be Strong Sign. Planets other than the above do not lend strength by aspect.

Lagna Sign: The Rising sign, at a particular time, is the sign of the zodiac positioned on the eastern horizon on the cusp of the first house at birth and is called Lagna. The Lagna lord is Atmakarka and is considered more powerful in Lagna.

Badhaka Signs: Similarly, there are Badhaka Rashi defined as per the Janma-Lagna. It works as the Badhaka Rashi and harms badly that house in which it is occupying.

Gandanta and its Effects: The ending portions, 30th degree of Cancer, Scorpio and Pisces are called Gandanta. It is said, that one born in Gandanta will not survive. He will either lose his mother, or he will end the dynasty, i.e. he is the last of his descent and will not have any children. If, however, he survives, he becomes a king with many elephants and horses.

4.2 Sign Characteristics

Zodiac has twelve signs having different characteristics. They act in different ways and are also known for its different nature. These are given in the following tables:

Table 1: Sign Characteristics

Ascendant	Benefic Planets	Malefic Planets	Most Malefic	Neutral Planets
Aries	Mars, Sun, Jupiter,	Venus, Saturn	Mercury	Moon
Taurus	Venus, Sun, Mars, Mercury, Saturn	Moon	Jupiter	--
Gemini	Mercury, Venus, Saturn	Sun, Jupiter	Mars	Moon
Cancer	Moon, Mars, Jupiter	Mercury, Venus	Saturn	Sun
Leo	Sun, Mars, Jupiter	Moon, Mercury, Venus	Saturn	--
Virgo	Mercury, Venus, Saturn	Sun, Moon, Jupiter	Mars	--
Libra	Venus, Mercury, Saturn	Sun, Moon	Jupiter	Mars
Scorpio	Mars, Sun, Moon, Jupiter	Venus	Mercury	Saturn
Sagittarius	Jupiter, Sun, Mars	Moon, Mercury, Saturn	Venus,	--
Capricorn	Saturn, Mercury, Venus	Moon, Mars, Jupiter	Sun	--

Aquarius	Saturn, Sun, Mars, Venus	Moon, Mercury	Jupiter	--
Pisces	Jupiter, Moon, Mars	Sun, Mercury, Saturn	Venus	--

Table 2: Sign Characteristics

Lagna (Ascendant) Sign	Death inflictor (Maraka Graha)	Raja Yoga Karaka Graha	Neutral Planets	Badhaka Rashi (Signs)
Aries (Mesha)	Mercury, Venus	Jupiter	Venus	Aquarius
Taurus (Vrishabha)	Jupiter, Venus, Moon	Saturn	Sun	Scorpio
Gemini (Mithuna)	Mar, Sun, Moon	Saturn	Moon	Leo
Cancer (Karka)	Venus, Saturn, Mercury	Mars	Venus	Taurus
Leo (Simha)	Saturn, Venus, Moon	Mars	Mercury	Aquarius
Virgo (Kanya)	Jupiter, Moon, Sun	Venus	Moon	Scorpio
Libra (Tula)	Jupiter, Sun	Mercury, Rahu	Mars	Leo
Scorpio (Vrishchika)	Mercury, Venus, Saturn	Moon	Saturn,	Taurus
Sagittarius (Dhanu)	Venus, Moon, Mercury	Sun, Ketu	Mercury	Aquarius
Capricorn (Makar)	Mars, Jupiter	Mercury	Mars	Scorpio
Aquarius (Kumbh)	Sun, Jupiter,	Venus	Sun	Leo

	Moon			
Pisces (Meena)	Saturn, Venus, Sun, Mercury	Mars	Mercury	Taurus

Note: In case of following Lagna, some planets act as special death inflictor:

Mesh Lagna: Sani will also inflict death, if associated with an adverse Graha.

Vrishabha Lagna: Mangal will inflict death, if associated with an adverse Graha.

Mithuna & Simha Lagna: Chandra is the prime killer.

Karka Lagna: Surya will also inflict death, if associated with an adverse Graha.

Kanya Lagna: Sukra Yuti with Buddha will produce Raja Yoga. Sukra will also inflict death, if associated with an adverse Graha.

Tula Lagna: Guru will also inflict death, if associated with an adverse Graha.

Vrishchika Lagna: Sukra will also inflict death, if associated with an adverse Graha.

Dhanu Lagna: Sani and Sukra will also inflict death, if associated with an adverse Graha.

Makar Lagna: Sani and Mangal will also inflict death, if associated with an adverse Graha. Sukra is capable of causing a superior Yoga.

Table 3: Sign Characteristics

Lagna Sign	Person's Nature (Guna)	Cause of Death	Person's Nature	Varna
Aries (Mesha)	Rajasic	High Fever	Violent	Kshatriya (Warrior)
Taurus (Vrishabha)	Rajasic	Fire, Weapon	Auspicious	Sudra (Service Person)
Gemini (Mithuna)	Rajasic	Cataract, Asthma, Mental Deviation,	Violent	Vysya (Trader)

		Loss of Appetite		
Cancer (Karka)	Sathwic	Cholera	Auspicious	Brahmana (Intellectual)
Leo (Simha)	Sathwic	Wild Beast, Fever, Boils, Enemies	Violent	Kshatriya (Warrior)
Virgo (Kanya)	Thtamasic	Women, Venereal Disease, Fall from height	Auspicious	Sudra (Service Person)
Libra (Tula)	Sathwic	Brain Fever, Typhoid	Violent	Vysya (Trader)
Scorpio (Vrishchika)	Sathwic	Jaundice	Auspicious	Brahmana (Intellectual)
Sagittarius (Dhanu)	Sathwic	Tree, Water, Wood, Weapon	Violent	Kshatriya (Warrior)
Capricorn (Makar)	Thamasic	Stomach Ache, Loss of Appetite	Auspicious	Sudra (Service Person)
Aquarius (Kumbh)	Thamasic	Cough, Fever, Consumption	Violent	Vysya (Trader)
Pisces (Meena)	Sathwic	Drowning	Auspicious	Brahmana (Intellectual)

Table 4: Sign Characteristics

Lagna Sign	Sign Lord	Affecting Disease	Affected Part of Body

Aries (Mesha)	Mars	Bile (Pitta)	Head, Face, & Brain
Taurus (Vrishabha)	Venus	Cold (Sleshma)	Neck, Throat, & Gland
Gemini (Mithuna)	Mercury Rahu	Gas (Vata)	Shoulder, Lungs, Hand, Blood, & Hand's Bone
Cancer (Karka)	Moon	Cold	Chest, Breast, Stomach, Shoulder's Bones,
Leo (Simha)	Sun	Bile	Back, Waist, Heart, Spinal Bones
Virgo (Kanya)	Mercury	Gas	Lever, Back Bones, Tili, Gurda
Libra (Tula)	Venus	Cold	Skins
Scorpio (Vrishchika)	Mars Pluto	Bile	Penis, Thighs, Nectar
Sagittarius (Dhanu)	Jupiter Ketu	Gas	Waist, Veins
Capricorn (Makar)	Saturn	Gas	Knees, Bone's joints
Aquarius (Kumbh)	Saturn Uranus	Gas	Feet, Digestive Organs
Pisces (Meena)	Jupiter	Gas	Ankles, Palms

Table 5: Sign Characteristics

Lagna Sign	Aspect on other Signs	Sign Age	Relation
Aries (Mesha)	Leo, Scorpio Aquarius	28 ½ YEARS	Chandra Sadga

Taurus (Vrishabha)	Libra, Cancer Capricorn	18 YEARS	Chandra Sadga
Gemini (Mithuna)	Sagittarius, Virgo Pisces	33 ½ YEARS	Chandra Sadga
Cancer (Karka)	Taurus, Libra Capricorn	40 YEARS	Chandra Sadga
Leo (Simha)	Aries, Scorpio Aquarius	28 ½ YEARS	Surya Sadga
Virgo (Kanya)	Gemini, Sagittarius Pisces	18 YEARS	Surya Sadga
Libra (Tula)	Taurus, Cancer Capricorn	33 ½ YEARS	Surya Sadga
Scorpio (Vrishchika)	Aries, Aquarius Leo	40 YEARS	Surya Sadga
Sagittarius (Dhanu)	Gemini, Virgo Pisces	28 ½ YEARS	Surya Sadga
Capricorn (Makar)	Aspect on other Signs	18 YEARS	Surya Sadga
Aquarius (Kumbh)	Leo, Scorpio Aquarius	33 ½ YEARS	Chandra Sadga
Pisces (Meena)	Libra, Cancer Capricorn	40 YEARS	Chandra Sadga

Table 6: Sign Characteristics

Lagna Sign	Sign Gender	Mode	Sign Element (Tatva)	Affected Part of Body
Aries (Mesha)	Male	Odd	Fire	Head, Face, Brain
Taurus (Vrishabha)	Female	Even	Earth	Throat, Gland Right Eye
Gemini (Mithuna)	Male	Odd Dual	Air	Neck, Nose, Lungs, Blood, Hand's Bone Ear,

Cancer (Karka)	Female	Even	Water	Chest, Breast, Stomach, Shoulder's Bones,
Leo (Simha)	Male	Odd	Fire	Upper Stomach, Back, Waist, Heart, Spinal Bones
Virgo (Kanya)	Female	Even Dual	Earth	Digestive Organs, Kidney, Lever, Back Bones, Tili, Gurda
Libra (Tula)	Male	Odd	Air	Skins
Scorpio (Vrishchika)	Female	Even	Water	Uterus, Ovary, Penis, Nectar
Sagittarius (Dhanu)	Male	Odd Dual	Fire	Thighs, Waist, Veins
Capricorn (Makar)	Female	Even	Earth	Knees, Feet, Digestive Organs
Aquarius (Kumbh)	Male	Odd	Air	Calf Mussels, Feet, Digestive Organs
Pisces (Meena)	Female	Even Dual	Water	Left Eye, Ankles, Palms

Table 7: Sign Characteristics

Planet	Moola-Trikona Sign	Moola-Trikona. Degree	Detriment Sign
Sun	Leo	0 – 20	Aquarius
Moon	Taurus	4 – 30	Capricorn
Mars	Aries	0 – 12	Libra Taurus
Mercury	Virgo	16 – 20	Sagittarius Pisces

Jupiter	Sagittarius	0 - 10	Gemini Virgo
Venus	Libra	0 – 15	Scorpio
Saturn	Aquarius	0 - 20	Cancer Leo
Rahu	Virgo	--	--
Ketu	Pisces	--	--

Table 8: Sign Characteristics

Planet	Exaltation Sign	Debilitation Sign	Max. Exaltn. Debilitn. Degree	Detriment Sign
Sun	Aries	Libra	10	Aquarius
Moon	Taurus	Scorpio	3	Capricorn
Mars	Capricorn	Cancer	28	Libra Taurus
Mercury	Virgo	Pisces	15	Sagittarius Pisces
Jupiter	Cancer	Capricorn	5	Gemini Virgo
Venus	Pisces	Virgo	27	Scorpio
Saturn	Libra	Aries	20	Cancer Leo
Rahu	Taurus Gemini	Scorpio Sagittarius	--	--
Ketu	Scorpio Sagittarius	Taurus Gemini	--	--

Table 9: Sign Characteristics

Lagna Sign	Element	Mode of Expression	Positive Quality	Negative Quality
Aries	Fire	Cardinal	Vital	Impulsive
Taurus	Earth	Fixed	Stable	Stubborn
Gemini	Air	Mutable (Dual Signs)	Adaptable	Cursory
Cancer	Water	Cardinal	Protective	Jealous
Leo	Fire	Fixed	Authority	Autocratic

Virgo	Earth	Mutable (Dual Signs)	Detailed	Critical
Libra	Air	Cardinal	Diplomatic	Vacillating
Scorpio	Water	Fixed	Resurgent	Ruthless
Sagittarius	Fire	Mutable (Dual Signs)	Discerning	Moralistic
Capricorn	Earth	Cardinal	Principled	Miserly
Aquarius	Air	Fixed	Liberal	Eccentric
Pisces	Water	Mutable (Dual Signs)	Charity	Anxiety

Table 10: Sign Characteristics

Lagna Sign	Person's Characteristics	Person's Profession
Aries (Mesha)	Hasty, impulsive, restless, short-tempered	Govt. job, surgeon, mechanics, industrialists, athletes, Police, Military Service, Fire Service, Sports, Engineering, arm manufacturing, trade union leader
Taurus (Vrishabha)	Slow in movement, inclined to ease and luxury, faithful & obedient	Musician, singer, actors, banking, tailors, property dealing, Jewellery business, money lending, commission agent, financial institutions, handicrafts, fancy articles, scented materials, five star hotels, drama, cinema, music, poet, story writer.
Gemini (Mithuna)	Good speakers, witty and humorous, inquiring and curious, fond of knowledge,	media and journalism, accountants, translators, writers, Information and broad casting, space department, education department, book publishing , mathematics department, auditors, law and

	fun seeking,	order councillor, ambassador.
Cancer (Karka)	Emotional , forgiving, sensitive	Export and Import, naval and marine, fishing, nursing, interior design, food, petroleum, historians, shipping, transport department, agriculture, hotel business.
Leo (Simha)	Dominative, behaves like ruler	Govt. Job, Politics, Administrator, Social Services, Charitable institutions, Engineering, Industry, religion, investing, diplomacy.
Virgo (Kanya)	Intelligent, good speaker, tactful	Auditing, Accounting, Business, Teacher, writer, retail shops, computing, astrology, media, doctors, healing.
Libra (Tula)	Good talker, judicious in dealings	Shop, commission agents, bank, Life insurance, law department, hotel business, bar and Restaurant, Dancing Hall, Beauty parlour, Music, Dance , Cinema, judges, artists, cosmetics, fashion, receptionists, advertising, interior decorating, prostitutes.
Scorpio (Vrishchika)	Peevish, straight forward, likes to hide or run away from people and crowds	Iron Industries, Engineering, and Instrument Manufacturing, raw materials, priest, astrology, mantra and tantra, occult practices, chemicals, drugs, liquids, insurance, doctors, nurses, police, occult.
Sagittarius (Dhanu)	Honest, easy going, even-tempered, kind hearted, gambles	Forest department, law, religion, banking and finance, entrepreneurs, athletes, law department, temple, financial institutions, education department, ordnance depot, military training, social service, charitable institutions.

Capricorn (Makar)	Witty, and changeable, good organizer, cautious, secretive, ambitious, preserving, pragmatic	Hotels, food products, engineer, doctor, business, building work, Granite stone and sand business, Labourer like porters, coolies, drivers, shoe polishing, shoe makers, plumber and mining.
Aquarius (Kumbh)	Studious, philosopher, honest, benevolent	advisors, consultants, philosophers, astrologers, engineers, computer, Psychology, Religion, Teaching, Research and Development, Administration, Service in Space Dept., Defence, Fire, Jail, Bomb manufacturing, tourist guide, central excise CBI Dept.
Pisces (Meena)	Lazy, emotional, timid, honest, talkative, intuitive , psychic, fond of good food and company	doctors, captain, hospital, prisons Education Department, Religious Institutions, Medicine, Financial, Law Department, External Affairs, Bank, Navy, shipping, temple worker, priest

Table 11: Sign Characteristics

Name of Sign	Effects	Direction	Progeny Nature	Feature
Aries	Positive	Northern	Barren	Bestial
Taurus	Negative	Northern	Semi-Fruitful	Bestial
Gemini	Positive	Northern	Barren	Human
Cancer	Negative	Northern	Fruitful	Bestial
Leo	Positive	Northern	Barren	Bestial
Virgo	Negative	Northern	Barren	Human
Libra	Positive	Southern	Semi-Fruitful	Human

Scorpio	Negative	Southern	Fruitful	Bestial
Sagittarius	Positive	Southern	Semi-Fruitful	Half Bestial & Human
Capricorn	Negative	Southern	Semi-Fruitful	Bestial
Aquarius	Positive	Southern	Semi-Fruitful	Human
Pisces	Negative	Southern	Fruitful	Half Bestial & Human

Table 12: Sign Characteristics

Lagna Sign	Element	Mode of Expression	Positive Quality	Negative Quality	Affected Part of Body
Aries	Fire	Cardinal	Vital	Impulsive	Head, Face, & Brain
Taurus	Earth	Fixed	Stable	Stubborn	Neck, Throat, & Gland
Gemini	Air	Mutable	Adaptable	Cursory	Shoulder, Lungs, Hand, Blood, & Hand's Bone
Cancer	Water	Cardinal	Protective	Jealous	Chest, Breast, Stomach, Shoulder's Bones,
Leo	Fire	Fixed	Authority	Autocratic	Back, Waist,

					Heart, Spinal Bones
Virgo	Earth	Mutable	Detailed	Critical	Lever, Back Bones, Tili, Gurda
Libra	Air	Cardinal	Diplomatic	Vacillating	Skins
Scorpio	Water	Fixed	Resurgent	Ruthless	Penis, Thighs, Nectar
Sagittarius	Fire	Mutable	Discerning	Moralistic	Waist, Veins
Capricorn	Earth	Cardinal	Principled	Miserly	Knees, Bone's joints
Aquarius	Air	Fixed	Liberal	Eccentric	Feet, Digestive Organs
Pisces	Water	Mutable	Charity	Anxiety	Ankles, Palms

Table 13: Sign Characteristics

Lagna Sign	Person's Characteristics	Person's Profession
Aries (Mesha)	Hasty, impulsive, restless, short-tempered	Govt. job, surgeon, mechanics, industrialists, athletes, Police, Military Service, Fire Service, Sports, Engineering, arm manufacturing, trade union leader
Taurus	Slow in	Musician, singer,

(Vrishabha)	movement, inclined to ease and luxury, faithful & obedient	actors, banking, tailors, property dealing, Jewellery business, money lending, commission agent, financial institutions, handicrafts, fancy articles, scented materials, five star hotels, drama, cinema, music, poet, story writer.
Gemini (Mithuna)	Good speakers, witty and humorous, inquiring and curious, fond of knowledge, fun seeking,	media and journalism, accountants, translators, writers, Information and broad casting, space department, education department, book publishing , mathematics department, auditors, law and order councillor, ambassador.
Cancer (Karka)	Emotional , forgiving, sensitive	Export and Import, naval and marine, fishing, nursing, interior design, food, petroleum, historians, shipping, transport department, agriculture, hotel business.
Leo (Simha)	Dominative, behaves like ruler	Govt. Job, Politics, Administrator, Social Services, Charitable institutions, Engineering, Industry, religion, investing, diplomacy.
Virgo (Kanya)	Intelligent, good speaker, tactful	Auditing, Accounting, Business, Teacher, writer, retail shops,

		computing, astrology, media, doctors, healing.
Libra (Tula)	Good talker, judicious in dealings	Shop, commission agents, bank, Life insurance, law department, hotel business, bar and Restaurant, Dancing Hall, Beauty parlour, Music, Dance , Cinema, judges, artists, cosmetics, fashion, receptionists, advertising, interior decorating, prostitutes.
Scorpio (Vrishchika)	Peevish, straight forward, likes to hide or run away from people and crowds	Iron Industries, Engineering, and Instrument Manufacturing, raw materials, priest, astrology, mantra and tantra, occult practices, chemicals, drugs, liquids, insurance, doctors, nurses, police, occult.
Sagittarius (Dhanu)	Honest, easy going, even-tempered, kind hearted, gambles	Forest department, law, religion, banking and finance, entrepreneurs, athletes, law department, temple, financial institutions, education department, ordnance depot, military training, social service, charitable institutions.
Capricorn (Makar)	Witty, and changeable, good organizer, cautious, secretive, ambitious,	Hotels, food products, engineer, doctor, business, building work, Granite stone and sand business, Labourer like porters, coolies, drivers,

	preserving, pragmatic	shoe polishing, shoe makers, plumber and mining.
Aquarius (Kumbh)	Studious, philosopher, honest, benevolent	advisors, consultants, philosophers, astrologers, engineers, computer, Psychology, Religion, Teaching, Research and Development, Administration, Service in Space Dept., Defence, Fire, Jail, Bomb manufacturing, tourist guide, central excise CBI Dept.
Pisces (Meena)	Lazy, emotional, timid, honest, talkative, intuitive, psychic, fond of good food and company	doctors, captain, hospital, prisons Education Department, Religious Institutions, Medicine, Financial, Law Department, External Affairs, Bank, Navy, shipping, temple worker, priest

4.3 Synastry by Sign Element

Every Zodiac Sign falls into one of four elements. There are four Elements, such as earth represents common sense; fire represents action, air represents thinking and communication skills, and water represents the ability to feel and intuitively know. Each Element is assigned to each sign depending to their orientation in the zodiac. Many astrologers consider the element of each of the planets when determining which of the elements may be more significant in a horoscope.

Fire (Aries, Leo, and Sagittarius): Fire is active and masculine. People of the Fire element are outgoing, quite moral, very creative, courageous, passionate, impulsive, hot, dynamic, progressive, action oriented, and direct. Their essence is spirit. They are enthusiastic, optimistic, confident, naive, self-centred, open, confronting, loyal, tactless, impatient, honest, trusting, and independent and feel free.

Earth (Taurus, Virgo, and Capricorn): Earth is a receptive, feminine sign. People are practical, cautious, and pragmatic approach to life and build solid, 'real' material success, i. e. car, home, career success and have long range planning and strong determination to succeed. They are safe/secure, suspicious, sensual, organised, dependable, introvert, and efficient and strong survival instinct.

Air (Gemini, Libra, and Aquarius): They are active, curious, idealistic, unemotional, conceptual, devoid of feeling, good to communicate, social, objective, impersonal, distant, masculine, intellectual, changeable, and impractical, good speech, and natural communicators, extroverted, social, charming, and logical and air has least obvious bad qualities, theoretical, abstract, needs to socialise, needs to share ideas.. The lack of Air Element in a native birth chart indicates difficulty in the expression of that person. Communication of ideas and the ability to conceptualise may prove difficult.

Water (Cancer, Scorpio, and Pisces): People of Water element dissolve everything in them coolly. They take the shape of who they are with, and are quite emotional, sustaining, emotional, sensitive, imaginative, protecting, compassionate, caring, artistic, moody, soulful, subconscious, irrational, introverted, but strong/powerful, vulnerable to hurt, intimate, defensive, psychic, past, suffering, suspicious initially, self-contained, picks up impressions and associated with healing.

Calculation of Element Strength: My method for evaluating the strength of an element in a birth chart is to assign a value of 4 to the element associated with the Sun; the Moon element is assigned a value of 3; Mercury, Venus, and Mars sign elements are assigned a value of 2 each, and Jupiter and Saturn each have a value of 1. Uranus, Neptune and Pluto are disregarded because their element is more societal affecting large groups of individuals born during a period. Using this approach, if as many as 8 points are concentrated in one element, it is considered "Preponderance" in that element. If we get less than 8 points with this approach in one element, it is considered "Absence" in that element.

Example: In one chart, the planets are placed as is given below: For Sun in Aquarius, an Air sign, we assign = 4 Points.

For Moon in Libra, an Air sign, we assign = 3 Points. For Mercury in Aquarius, an Air sign, we assign = 2 Points. For Venus in Capricorn, an Earth sign, we assign = 2 Points. For Mars in Sagittarius, a Fire sign, we assign = 2 Points. For Jupiter in Pisces, a Water sign, we assign = 1 Points. For Saturn in Fire, a cardinal sign, we assign = 1 Points.

Thus, we have Element strength, such as, 9 in the Air Element, 2 in the Earth Element s, and 3 in the Fire Element and 1 in Water Element in this chart. This shows a preponderance of Air element in this chart. The preponderance reading of Air element would be appropriate in the above chart.

The preponderance readings of all elements are given below:

A Preponderance of the Fire element: A preponderance of Fire Element indicates high spirits, great faith in self, enthusiasm, direct, honesty, intensely assertive, most daring, individualistic, active and self-expressive, good natured, fun loving, natural leader, having a good time than on material possessions, big egos. He believes so strongly in his own powers and abilities that he overlooks and frequently fails to take advantage of the talents and abilities of others. He tries to do it all himself and don't delegate well. He is constantly "out front" or "on stage", such as an Artist and they need to be recognized and admired for his attainment and accomplishments. Appreciation is more important than money in his estimation. Nothing hurts him more than being ignored. The fire sign sense of honesty is straightforward and often child-like. Thus, he believes everyone is, like himself, an open book.

A Preponderance of the Earth Element: A preponderance of Earth Element indicates cautious, conventional, dependable but quite responsible, methodical, organizer, a builder, and a hard-worker. It provides the skills and attitude necessary to succeed readily in the world of business and never gamble or take unnecessary chances. They understand the reality of a situation and value, reliable and steadfast. They are dependent, diligence and a pragmatic, no-nonsense approach to life. Lack of ideas or imagination, dullness, rigid, conservatism, extreme materialism, and blind adherence to rules and regulations are their potential faults.

A Preponderance of the Air element: The preponderance of Air Element suggests a strong emphasis on thought, ideas and intellectual and they communicate and express ideas with mental agility and become the impractical dreamers, constantly thinking, people-oriented, but more inclined toward the group than the individual. Your interests are varied, and you're apt to be a life-long student.

A Preponderance of the Water element: The preponderance of Water Element indicates close emotional relationships, romantic, sentimental, affectionate, secure bond with partner, communicate best in non-verbal ways; emotionally, psychically, or through forms as art, dance music, poetry and photography. They have a natural feel and sense for the arts and are apt to let the heart rule the head, highly impractical and impressionable.

4.4 Synastry by Ascendant (Lagna):

General Predictions: If the person is born in the Mesh, he/she is brave and a thief; in Vrishabha wealthy, Mithuna learned; Karka king; Simha respected by king; Kanya learned; Tula minister or adviser; Dhanu sinful; Kumbh businessman; and in Meena he/she is wealthy.

Prediction by Sign Ascendant, 1st House (Thanu/ Body):

Aries (Mesha): He/She is proud, wealthy, having excessive anger, dependent on others. He/She will be medium height, white complexion, smiling face, clever, and lean. He/She suffers from abdominal problems. He/She relishes helping the poor and has faith in God. He/She thinks very high but implements little. He/She inherits huge paternal property. He/She is prone to drowning and accidents. He/She is selfish and forgets a person who is no longer of use. He/She is prone to be cheated by friends and partners. He/She will improve in life and will earn money and believes in donation. Even born in medium income group, he/she earns good money of his/her own. He/She has a personality that is positive, aggressive, and competitive. He/She has leadership qualities, is bold, and empowered with more physical strength. He/She is best in sports, games, trekking, summer camp, and any other outdoor activities.

Taurus (Vrishabha): He/She is a pleasant talker, scholar and loves all. He/She is tall, luxurious, clean-hearted, strong built, good personality, whitish complexion, smiling face, clever and attractive personality. He/She improves in life and earns money and believes in donation. He/She is prone to Litigation or imprisonment because of personal or property disputes.

He/She is financially well off, earning much more. He/She is educated and enjoys a happy married life with educated and glorious progeny. He/She has differences with near relatives and is thus socially unpopular. He/She is prone to accidents. He/She has a tendency toward being heavy by both bone structure and the self-indulgence. Lord Krishna, Mata Amritanand Mayi and also Shri Basaveshwara were born in Taurus, but in Rohini Nakshatra. The second Drekana of Taurus gives the skill of fine arts, music and dance to the native.

Gemini (Mithuna): He/She is proud, loves his friends, charitable, and wealthy. He is annihilator of his enemies and progresses slowly in life. He/She will be medium height, whitish complexion and a faithful friend. He/She is sweet-voiced, jolly and humorous. He/She is well-wisher for everyone and gets cooperation from parents. He/She seldom seeks help and doesn't work under any one. He/She has his own successful business set-up - big or small. He/She is considerate to subordinate and weaker people. He/She has long life, lean body. He/She improves in life; earns money and believes in hard working. He/She, even, born in medium income group, earns money of his/he own. He/She does not get help from his/her spouse. 3rd Drekana of Gemini blesses him/her with fine art, music and dance.

Cancer (Karaka): He/She is religious, handsome, long, has good personality, whitish complexion, and donating. He/She is good-looking, rich and famous. He/She respects his elders and teachers. He/She is prone to head injury during childhood. He/She excels his business away from birth place. He/She is prone to be cheated by partners and close relatives. He/She gains from business abroad in white coloured items. He/She is intelligent and heads an organisation or society. People flock to him/her for advice. He/She may undergo political imprisonment. He/She, even, born in medium income group, earns money of his own. He/She does not enjoy his life due to hardship. His family life is not happy and always difference of opinions between husband and wife. He/She is very protective of those who are close to him/her. He/She is affectionate, emotional, home loving and lovely in their approach.

Leo (Simha): He/She is annihilator of his enemies, has few children, is tall, strong built, good personality, whitish complexion, and smiling face, clever and has attractive personality. He/She is efficient, undertakes tough tasks, and is hard-hearted and always successful. He/She is self-dependent and doesn't trust others. He/She spends as quickly as he earns. He/She crushes his/her enemy, is religious and donating. He/She doesn't forget or forgive his enemy and takes revenge. He/She has differences with father. His wife is long-lived but he/she keeps quarrelling with his/her. He/She is a devoted friend who will remember and repay a kindness. He/She has royal tastes and a sense of luxury. He/She is angry but vents the anger quickly. He/She goes to any authority to prove that he/she is right.

Virgo (Kanya): He/She is endowed with beauty and has a good fortune, medium height, broad chest, whitish complexion, smiling face, clever, very fast in doing the job and is selfish and harm too much out of his selfishness. His/Her young age is very happy. He/She is a successful in politics because he/she has something inside and speaks something else in public. Nobody can measure his political capability. He/She is never crude or coarse. He/She prefers the role of researcher, observer, critic, or teacher. He/She is fault finding type and hypercritical. He/She feels proud in finding the fault and drawback in others.

Libra (Tula): He/She is a scholar, earns his livelihood by virtuous means, wealthy and is respected by everybody. He/She is tall, fair complexion and healthy. He/She is sweet-voiced and benevolent. He/She doesn't stick to one profession and keeps spending too much on research. He/She enjoys little reputation at home but is reputed outside. He/She excels in occupations related to iron. He/She has problem with brothers/sisters. He/She is talkative, is not hard-working and depends on fate, and thus leads insecure life. He/She is unlucky for father's business and gives setback at the age of 12. Initially he/she begins with service but later settles down in own business. He/She leads happy married life with kids.

Scorpio (Vrishchika): He/She is wealthy and a scholar. He/She is tall, lean, either very rich or very poor. He/She lives away from home since very early in life and dominates over his family members as well as outsiders. He/She doesn't forget or forgive his enemy. He/She has his own business. He/She has financial problems up to 30 years but later he/she earns money and supports others. He/She cannot sit idle and is world-famous and heads an organisation or society. He/She is prone to injury during fight or accident. He/She gets married more than once. He/She will be rich, famous, popular and smart in love matter. He/She is successful in politics. He/She earns sufficient money in the life but not much savings. This Sign is not good for domestic happiness.

Sagittarius (Dhanu): He/She is an expert in policy matters, religious, important person in his family, medium height, whitish complexion, smiling face, and attractive personality. He/She has strong faith in God, is vegetarian, simple living, believes the people very easily, and is a businessman. He/She does his work with well planning. He/She likes discipline, truth, justice, kindness and independence in his/her life. He/She takes everything and everyone for granted. His/Her reasoning powers are superb. He/She enjoys the good fortune of having thought patterns that remain young and fresh throughout life. He/She makes lots of promises but fail to maintain them. He/She has a great sense of fairness and adopts only fair means to handle the job.

Capricorn (Makar): He/She is inclined towards evil deeds, is greedy and has many children, but hard working. He/She is tall, strong built, good personality, whitish complexion, clever and but selfish, changing his faces frequently as per situations, and very talkative. He/She works under someone and subjects to heavy ups & downs. He/She spends immediately what he/she earns in bad deeds. He/She dominates spouse and quarrels with other family members on that account. He/She stays away without information. He/She is abusive and short-tempered. He/She fails in business and has to go in for service. He/She is attached to mother and has differences with father. He/She is helpful to brothers and sisters and has more daughters. His/Her expense is more than earning and hence always faces shortage of money. The conjugal life is not happy and there is always difference of opinions between them.

Aquarius (Kumbh): He/She leads a happy and contented life. He/She will be well educated, gentle, peaceful, always ready to help others, having good thinking; tall, whitish complexion and attractive personality and straight forward in nature. He/She successfully tackles early age problems and heads for good time later. He/She would spend any amount of time and money to crush his/her enemy or achieve his/her aim. He/She likes to gossip and interact with women and has interest in astrology. He/She is financially well off and is prone to chest problems.

He/She will be very hard working and faces difficulties in life. He/She may suffer with the stomach and heart diseases in old age. He/She is strong willed, detached and unyielding in nature. The child of this Sign is unpredictable regarding his/her behaviour and can change frequently himself/herself to any extent during his childhood.

Pisces (Meena): He/She is wealthy, educated, gentle, peaceful, religious, and always ready to help relatives, medium height, and beautiful curly hair and have self confidence. He/She is famous and heads an organisation or society. He/She studies very hard, but he/she is not a high scorer. He/She helps friends and serves society physically but without spending money. He/She works overtime to finish the work same day. He/She becomes favourable of family members and outsiders. He/She just cannot work under any one and leave his/her service very soon for own business. He/She loses temper beyond control but calms down very quickly. He/She is prone to cheating by partners and should better work alone. He/She is likely to break first marital relation (matured) or otherwise, be unhappy with spouse but happy with progeny. He/She may have great interest in writing, music and acting. He/She is likely to set high goals. Drug, alcohol and false promises attract him/her easily.

5

Planet (Graha)

5.1 Planets' Description

There are nine Graha, namely, Sun, Moon, Mars, Mercury, Jupiter, Venus, Saturn Rahu and Ketu. But after further studies, the Uranus, Neptune and Pluto have been added to the astrology to make it more fascinating subject. Moon, Mercury, Jupiter and Venus are benefic by nature and others are malefic.

Surya: Surya's eyes are honey-coloured. He has a square body. He is of clean habits, bilious, intelligent and has limited hair (on his head).

Chandra: Chandra is very windy and phlegmatic. She is learned and has a round body. She has auspicious looks and sweet speech, is fickle-minded and very lustful.

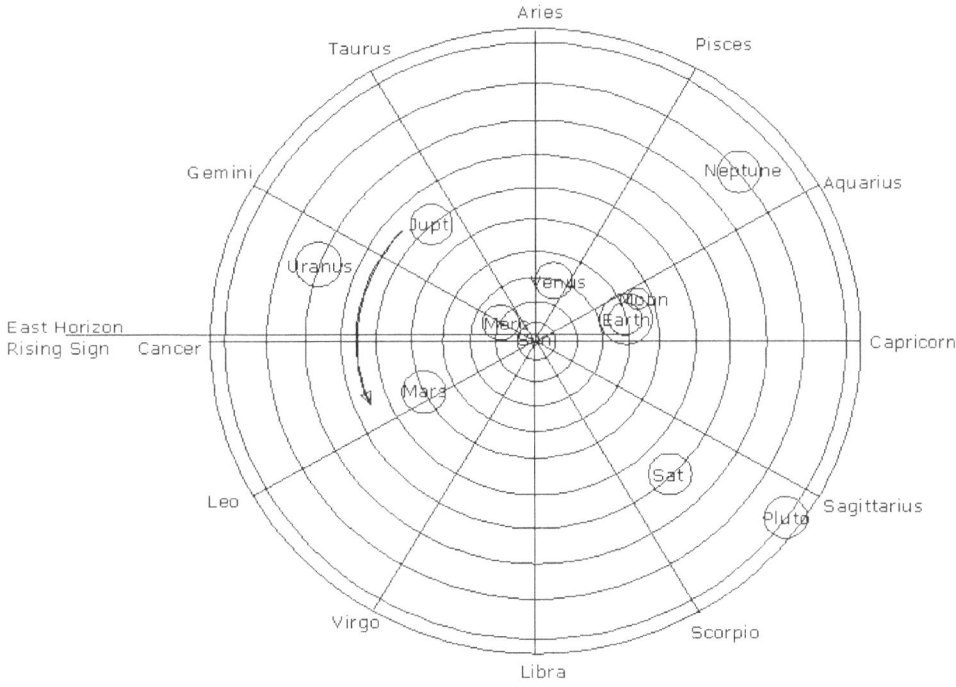

Fig: Planets position with respect to Rising Sign (Lagna) as View from the Earth

Mangal: Mangal has blood-red eyes, is fickle-minded, liberal, and bilious, given to anger and has thin waist and thin physique.

Buddha: Buddha is endowed with an attractive physique and the capacity to use words with many meanings. He is fond of jokes. He has a mix of all the three humours.

Guru: Guru has a big body, tawny hair and tawny eyes, is phlegmatic, intelligent and learned in Shashtra.

Sukra: Sukra is charming, has a splendours physique, is excellent or great in disposition, has charming eyes, is a poet, is phlegmatic and windy and has curly hair.

Sani: Sani has an emaciated and long physique, has tawny eyes, is windy in temperament, has big teeth, is indolent and lame and has coarse hair.

Rahu (The Dragon Head/ North Node of Moon): Rahu is the dead body of the lusty demon, killed by Vishnu. So, Rahu is the head of the dragon and symbolizes the trouble. Rahu has smoky appearance with a blue mix physique. He resides in forests and is horrible. He is windy in temperament and is intelligent.

Ketu (Dragons Tail/South Node of Moon): Ketu is the dead body of the lusty demon, killed by Vishnu. So, Ketu is the tail of the dragon and losses and symbolizes the death along with Saturn. Ketu has smoky appearance with a blue mix physique. He resides in forests and is horrible. He is windy in temperament and is intelligent. Ketu is akin to Rahu.

Note: Each planet has been allotted some points, such as Sun - 48; Moon - 49; Mars - 39; Mercury - 54; Jupiter - 56; Venus - 52; Saturn - 39. Thus it is totalling 337 points all together.

5.2 Planets' Characteristics

Table 1: Planet's Characteristics

Planets	Friends	Neutrals	Enemies
Sun (Surya)	Moon, Mars, Jupiter	Mercury	Saturn, Venus,
Moon (Chandra)	Sun, Mercury	Mars, Jupiter, Venus, Saturn	--
Mars	Sun, Moon,	Venus,	Mercury

(Mangal)	Jupiter	Saturn,	
Mercury (Buddha)	Sun, Venus,	Mars, Jupiter, Saturn	Moon
Jupiter (Guru)	Sun, Moon, Mars,	Saturn,	Mercury, Venus
Venus (Sukra)	Mercury; Saturn,	Mars, Jupiter	Sun, Moon
Saturn (Sani)	Mercury, Venus,	Jupiter,	Sun, Moon, Mars

Table 2: Plant's Characteristics

Planet	Sun	Moon	Mars	Mercury
Planet's Natural behaviour	Malefic	Benefic	Malefic	Benefic
Karaka	Father, Soul, and 1st house	Mind, Mother and 4th house	Younger brother and Husband.	Speech, education, & 3rd house.
Colour	Red-orange	Tawny	Red	Green
Cabinet	King	Queen	General	Prince
Deities	Agni	Varuna	Kartikkeya	Maha Vishnu
Sex	Male	Female	Male	Female
Tatva	Agni (fire)	Jala (water)	Agni (fire)	Bhumi (earth)
Varna (Deeds)	Kchatriya	Vaishhya (trader)	Kchatriya	Vaishya (trader)
Guna (Behaviour)	Sattva	Sattva	Tamas (ignorance)	Rajas (passion)
Dhatu (Body Part)	Asthi (bones)	Rakta (blood)	Majja (marrow)	Tvak (skin)
Time periods	Ýyana (half Yr,)	Kchana (second)	Vara (day)	Ritu (season)
Taste	Pungent	Saline	Bitter	Mixed
Ritu (Seasons)	--	Varsha (rainy)	Grishma (summer)	Sarad (Winter)
Lord	Lord Siva	Goddess	Lord	Maha

		Parvathi	Karthikeya	Vishnu
Dasa Period	6 years	10 years	7 years	17 years

Table 3: Planet's Characteristics

Planet	Jupiter	Venus	Saturn
Planet's Nature	Benefic	Benefic	Malefic
Karaka	Knowledge, happiness, Male, 5th house; & 9th house.	wife, Sex, Love, Younger sister, 2nd house, & 7th house.	Misery & Grief, Profession, Elder brother and 10th house
Colour	Tawny	Variegated	Black
Cabinet	Minister	Minister	Servant
Deities	Indra	Sukra	Brahma
Sex	Male	Female	Impotent
Tatva	Ýknna (ether)	Jala (water)	Vayu (air)
Varna (Deeds)	Brahmana (priest)	Brahmana (priest)	Chhudra (worker)
Guna	Sattva	Rajas (passion)	Tamas (anger)
Dhatus Body Part	Vasa (fat)	Virya (semen)	Snayu (muscle)
Time periods	Masa (month)	Pakcha (fortnight)	Varsha (year)
Taste	Sweet	Sour	Astringent
Ritu (Seasons)	Hemanta (Dew)	Vasanta (spring)	Sisir (fall)
Lord	Lord Dakshina Murthi	Maha Lakshmi	Lord Yama
Dasa Period	16 years	20 years	19 years

5.3 Synastry by Planet

Synastry is the art of partnership Astrology. It is a fascinating and illuminating study of how individuals interact with one another. Synastry means "together with the stars". It is the art of analysing a partnership of two or more people, for instance. Synastry is a comparison between the horoscopes of two people in order to determine their likely compatibility and to have strong partnership.

Each individual is born with a personal birth chart, which is a map of the heavens for the moment they took their first breath. Some might say the birth chart has the effect of stamping, or imprinting, the energies of the planets and signs on an individual. Each and every one of us has all 10 planets and luminaries in our charts, but their positions by sign, house, and aspect are individual to each. When we interact with others, the individual energies of our natal charts form special relationships with their individual energies. The resulting interplay is as complex and unique as our own personalities.

Many of us are familiar with the study of Sun Sign Compatibility. Some will ask, for example, "Does a Leo get along with a Scorpio?" While these comparisons have some value, they are very general. Many other factors are involved when evaluating the compatibility of two people. Although Synastry is complex, we can turn to some especially useful methods of studying relationships that will help shed light on our interactions. Below are some valuable pointers. Turning to Astrology, and more specifically, Synastry, will help us find answers.

It **makes the Synastry by Position** of Planet in house, Sign, Positions of Mars (Mangal), Position of Venus (Shukra), and all the planets of two persons' Kundali and tells how easy it is for them to live or work together. Synastry is to concentrate and find out the personal relationships or those one-to-one relationships that involve the union of two people by partnership. It is the comparison of two or more natal - Charts in order to analyze the individuals involved. The 7th House shows a partner. The Partner is represented by 7th House, planets in the 7th House or ruler of the 7th and the sign on the 7th House. For Partnership, compatibility only decides their fate. A good Partnership should grow and develop gradually from understanding and not impulse, from true loyalty and not just sheer indulgence. If they are not compatible with each other, they cannot continue to be together, if they are together also, in future that may get stained any time. So before entering into a commitment like Partnership, we should know the compatibility that can be done only through Practical Astrology.

Astrological compatibility is the branch of astrology that studies relationships by comparing natal horoscopes. Relationships between planets, signs, and houses (sectors of the chart) are described in words and or often rated in numbers to show to what degree, how, and in what ways one person is compatible or incompatible with another. The assessment is based on the patterns and distributions of signs or planets in the respective and combined charts.

Horoscope matching is an art and it is the highest responsibility of an astrologer. The fundamental concept of matching horoscopes varies from astrologer to astrologer. But a real matching should be based on the horoscope or natal chart and should reflects the Partners nature, mental physical tendencies, conduct, qualities, future etc., It should not be only by star matching, and that is more common in India now, and it shows the failure of compatibility in many cases. In addition, to horoscope matching a host of other factors such as longevity of the individuals, character, poverty, body status, radical strengths and indications, planetary nature and afflictions, currency of major and minor periods, Marakas (death inflicting planets), placement of Mars in their nativity (Mangal Dosha), time of query, propitiations, auspicious time for marriage etc. should be examined to arrive at a suitable Partnership.

5.4 Synastry by Partner's Planets

The "Composite Horoscopes" or "Composite Chart" to find out the aspects formed by Sun, Moon, Mercury, Venus and Mars, is a favourite method for analysing partner's relationships. This is the comparing of the planetary aspects by placements in one chart. Select a particular planet from partner's horoscope and put it in the same Sign in native Horoscope as it was in partner's horoscope with different identification marks. Begin plotting the planet on native chart taking from partner's Chart. This is called the "Composite Horoscopes" or "Composite

Chart". For example, if the Sun, in native chart, is in Leo, and, in partner's Chart, is also in Leo, this indicates that his/her Sun is in same Sign (conjunction). This aspect is not good for partnership or relationship. Similarly, place the planet from both the charts on one Chart and then check out how these placements are related.

Sun

The Sun denotes authority. If the Sun in one chart is square or opposition to the Sun in the other partner's chart, there are chances of good attraction and compatibility between the two. A Sun in conjunction, sextile, or Trine with the Sun of the partner is not good because it indicates a strong egoistic feelings with the partner. A favourable Sun to Sun aspect are square or opposition, which can largely smooth out a relationship.

Sun in partner's 1st house: It shows a good deal of physical attraction, but it also shows competition and ego conflicts, and the spirit of competitiveness. The relationship (marriage) is usually found strong, mutually attractive and with good physical compatibility but business partnership will be competitive and egoistic, conflicting and with the spirit of competitiveness.

Sun in partner's 2nd house: It shows good cooperation in business and in the financial side of the partnership and good for the relationship (marriage).

Sun in partner's 3rd house: It suggests good intellectual exchange and communication, great compatibility in the communication of ideas and an interesting and always engaging relationship. Mutual interests in subjects such as education, books, planning, decision making and travel are a definite assets for the two.

Sun in partner's 4th house: It produces an intense interaction at a very deeply emotional or visceral level. In marital relationships, the partner tends to provide a base of operation for the expansion of self-expression.

Sun in partner's 5th house: It confers great potential for emotional and romantic attraction. Whatever is the relationship, they enjoy each other's company and sharing of pleasure-oriented social activities such as parties, games, sports, and entertainment.

Sun in partner's 6th house: It represents more or less an authority figure in the relationship. There is a 'care' or 'protective' management aspect associated and they are apt to take on a role of protector or caretaker in this relationship.

Sun in partner's 7th house: It shows a strong mutual attraction, which is very good for marriages and close personal friendships. There is sexual attraction, as well as a mutual enjoyment of each other's company.

Sun in partner's 8th house: It suggests a mutual need for security and a sexual attraction, since the eighth house is strongly associated with sex.

Sun in partner's 9th house: It denotes a role of dominating and authoritarian tendency, which can stir resentment in the relationship.

Sun in partner's 10th house: It often inspires him to greater effort in career or creative effort; lessen the burdens, and help the other overcome fears and limitations.

Sun in partner's 11th house: It suggests good relationship, usually strong friendship, with common goals and objectives and provides occasion for working together in group activities.

Sun in partner's 12th house: It suggests a well 'tuned in' to him and subconscious connection. They may be something of a 'help-mate' to each other, especially with regard to understanding life as a whole.

Moon

The Moon denotes emotions. If the Moon in one chart is square or opposed to the Moon in the partner's chart, there are chances of some conflicting ideas existing between them. A Moon in conjunction with the Sun of the partner is a very good sign in that the Moon person has a wonderful emotional connection with the partner. A favourable Sun/Moon aspect are conjunction, sextile, or Trine, which can largely smooth out a relationship. **Moon 1st house:** It denotes a close emotional link in family and domestic affairs, and is good for marriage partners. This close emotional tie makes both of them feel good to be in the other's company. There is an instinctive melting of temperament and moods seem to blend together.

Moon 2nd house: It shows romantic relationships and shows a joint enjoyment of luxuries or other sensuous pleasures, such as dining together.

Moon 3rd house: It suggests a fascination on the part of the Moon person with the manner in which the mind of the partner works and the way mental problems are tackled.

Moon 4th house: It makes for close relationships in family and domestic issues and has strong emotional connection and good relationship. There is a strong sense of feeling 'at home' together with each other.

Moon 5th house: It often indicates an emotional and highly romantic attraction, a romantic relationship, good domestic and family affairs and they play a parental role toward each other.

Moon 6th house: It shows the Moon native knack of taking care of the other. The Moon person has an instinctive understanding of the partner's needs and the willingness to fill his needs.

Moon 7th house: It denotes significant ups and downs in the affair, and there is a keen awareness of each other's moods and feelings. The Moon individual is apt to bring a good deal of support and understanding to the relationship.

Moon 8th house: It is good for business relationships and relates to the handling of joint funds in a partnership and in some cases, stimulates interest in occult or psychic phenomena in the partner. Emotional quarrels over money may erupt, especially if the cross-aspects are adverse.

Moon 9th house: It highlights acceptance of one another's religious, ethical, social, and moral values and understanding of the other person's philosophy of life and building on this mutual understanding.

Moon 10th house: It shows a natural understanding of what they want out of life and the emotional support to achieve these ends and a tendency to be brought together.

Moon 11th house: It is good for friendship and group associations, and share of feelings and emotional reactions, and spending a good deal of time together. There is a sense of kinship in this placement, and they tend to treat each other like family members and produces extremely fast friendship.

Moon 12th house: It produces a strong emotional and intuitive link, making them extremely aware of one another's moods and feelings and denotes sympathetic understanding, empathy and mutual compassion.

Mars

Mars plays major roles in romance and in the match making process and relates to romantic desires or lust. It determines sexually aggressiveness, raw passion or lack of it in nature. This is why the symbol of Mars is also the symbol for the male. He is often responsible for the sense of 'love at first sight' and is considered for a very strong sexual attraction. But, love at first sight and a very strong sexual attraction don't necessarily make a long and durable relationship. There will be most auspicious fit in between the partners, if, Mars and Venus are in the same sign or in the same element of the partner's Chart. The attachment is much stronger if the planets form a conjunction or Trine aspect in both Charts.

Mars 1st house: It indicates a strong energizing influence on the partner with competition in the relationship. This is a very good for the relationship. This combination shows a strong sexual attraction.

Mars 2nd house: It produces often conflict over money matters.

Mars 3rd house: It produces many arguments and criticism.

Mars 4th house: It produces a tendency to challenging attitudes and frequently prompts changes if the relationship succeeds.

Mars 5th house: It produces strong romantic attraction of a sexual nature. The Mars person may be very sexually demanding and impulsive in pleasure-oriented activities. There is apt to be a lot of competition associated with this relationship.

Mars 6th house: It denotes a caring and close relationship like a doctor/patient or a therapist/client.

Mars 7th house: It stimulates the other into greater self-expression whether in partnership or competition and is a dynamic, action-oriented partner reflecting a good deal of impulsive behaviour and produces considerable sexual attraction, but provides no guarantee of a lasting relationship by itself. It does provide an active and often interesting attachment.

Mars 8th house: It denotes a dynamic business, corporation or scientific relationship and challenges each other to self-improvement and development. There is also often a sexual attraction shown by this placement.

Mars 9th house: It produces annoyances and conflict and forces the other to subscribe to particular religious, philosophical, educational, moral, or cultural beliefs. This can show a relationship founded upon mutual goals related to religious or spiritual projects.

Mars 10th house: It shows strong interactions between the two a mutual professional interests and objectives. The Mars person may be attracted to this partner for the purpose of gaining professional advancement and status.

Mars 11th house: It denotes dynamic interaction with friends; this includes mutual participation in group and organizational activities. There is also mutual encouragement in the achievement of goals and objectives relating to political reforms, humanitarian causes, scientific research, or technological invention. This is an excellent placement for individuals working together on such enterprises.

Mars 12th house: It suggests a good deal of action and interchange between the two. They can work together in compatible ways dealing with charitable service, religion and hospitals.

Mercury

Mercury 1st house: It suggests a great deal of mental stimulation and mental appeal to the potential partner. If badly aspect, there may be many misunderstandings and arguments.

Mercury 2nd house: It is good for business and professional relationships and the person who's Mercury resides in the second house of the other is likely to provide that partner many money-making ideas.

Mercury 3rd house: It denotes mental compatibility, and suggests that the Mercury person stimulates the partner in areas related to communication, research and perhaps travel. This is an excellent aspect for teacher/student relationships.

Mercury 4th house: It suggests that the Mercury person is likely to make ideas relating to the handling of domestic and family affairs and may also bring about more of an incentive to travel.

Mercury 5th house: It brings ideas and opinions that tend to act as a spur to creative expression, artistic issues, or in the rearing of children.

Mercury 6th house: It shows that the Mercury individual brings ideas and opinions regarding health and diet into the relationship.

Mercury 7th house: It provides ideas that complement and round out the partner and a mental understanding that can be beneficial in marriage.

Mercury 8th house: It is a good comparative placement for two individuals engaged in business together, and is good for understanding the deeper needs and the unconscious motivations in a close relationship.

Mercury 9th house: It is an excellent position for two people desiring to share philosophical and intellectual pursuits. This is a particularly good placement in a teacher/student relationship. This aspect tends to stimulate interest in religion, education, philosophy, and travel.

Mercury 10th house: It produces ideas and communication that influences or relates to the career or personal status of the partner.

Mercury 11th house: It promises a friendship, usually relating to group or organizational activities. You may both have common humanitarian goals and interests. This aspect is helpful in good romantic relationships because it produces a solid basis for friendship and common interests.

Mercury 12th house: It denotes a mutual interest in the hidden or the occult, psychology, or perhaps even creative endeavours and special interest in the partner's attitude toward life.

Venus

Venus plays major roles in romance and in the match making process. It represents the love, romance and emotional response to love and a very strong sexual attraction. The most auspicious fit in between the partners would be for Mars and Venus to be in the same sign or in the same element. The attachment is much stronger if the planets form a conjunction or Trine aspect with the partner's Chart.

Venus 1st house: It shows identification with each other emotionally, a harmonious relationship and which results in a strong romantic attraction that can lead to marriage.

Venus 2nd house: It is more often associated with business and financial dealing with each other. There is compatibility when it comes to making money together.

Venus 3rd house: It denotes very harmonious communication and a chance for a good relationship because they are tactful, diplomatic, and considerate of each other.

Venus 4th house: It suggests a generally peaceful situation in the domestic scene, and brings harmony into the family life. A couple living together, would usually prefer to go out for dining, entertainment, or social activities and brings harmony into the family life. This is a favourable combination for marriage and a peaceful, harmonious home life, especially in the later years.

Venus 5th house: It is one of the strongest combinations for romantic and sexual attraction, love, pleasure and romantic fulfilment after marriage.

Venus 6th house: It often indicates social friendships established at work. They are able to work together harmoniously, sharing interests in diet, cooking, and arts and crafts, which is good for employer and employee relationship.

Venus 7th house: It suggests a strong romantic attraction and inclination to enjoy each other and interact harmoniously after marriage. They tend to like the same things and are very considerate of one another's wants and desires.

Venus 8th house: It produces a very strong sexual attraction and fascination that yields an especially significant physical combination and feelings are very deep-rooted.

Venus 9th house: It suggests a mutual view of philosophy, religion, law, or higher education, which establishes greater harmony and compatibility in the relationship in society.

Venus 10th house: It denotes a connection in the professional or business world, and the Venus person provides a sense of charm and diplomacy aiding the partner's career or status, which shows a tendency for the partner to place the Venus person on a 'pedestal'.

Venus 11th house: It is a favourable position for establishing social rapport and establishing common friends. Goals and objectives of both individuals may be similar and compatible.

Venus 12th house: It denotes close psychic and emotional link in the relationship. You are sympathetic and compassionate toward each other, and this may be the basis of the attraction.

5.5 Synastry by Planets' Aspects:

The aspects formed between the Sun, the Moon, and the other planets in one chart and the Sun, the Moon, and the other planets in the partner's chart show the detail of how smoothly this relationship will work. If the negative aspects appear in the Charts, the relationship is apt to be a major struggle. Kuja Dosha or Angaraka also refers to all Natural Malefic Planets arranged in order of high level of malefic, such as Saturn, Mars, Rahu, Ketu and Sun. Dosha means affliction or blemish for adverse effects and negative results. All malefic are giving adverse effects more or less. But in case of marriage, Mars are given more importance and its other's effects are ignored because it is negligible. The effect of affliction increases if the same is seen from Chandra Lagna and also from Venus or Jupiter.

Saturn Conjunction

Saturn conjunct Sun: The Sun-Saturn energy in synastry is much like the energy between a parent and child. A Saturn conjunction Sun aspect gives your partner soundness of judgment and emphasizes his/her ability to focus on what matters to you both and builds a strong sense of purpose.

Saturn conjunction Moon: Moon can be more vulnerable to Saturn's restrictive pressure. You will feel emotionally responsive to the ideas of partnerships.

Saturn conjunction Mars: This relationship can breed a lot of resentment if the Mars person isn't able to tolerate restrictions well. You are going to be feeling serious limits on your actions as your partner's caution and conservatism puts the brakes on your new ideas and ventures.

Saturn conjunction Mercury: Synastry between Saturn and Mercury can be excellent for a business or a mentor-student relationship. You will be more able to solve problems, work on serious issues, and focus on practical concerns because of your partner's influence, but you may feel inhibited about sharing thoughts and ideas which would not meet with approval.

Saturn conjunction Jupiter: This can be a highly productive relationship, as Saturn and Jupiter "get" one another and respect each other's necessary functions (stability and expansion.) You may feel like you have known each other forever, but your relationship is one which continues to grow over time. Your patience and steady dedication to your mutual goals will be rewarded substantially.

Saturn conjunction Venus: Venus in an Earth sign deals better with the responsibility and commitment demanded by Saturn. You have a serious desire to provide a secure foundation of loyalty, responsibility and commitment to your relationship and expect your partner to do the same.

Saturn conjunction Saturn: The key point is that you will strengthen each other's natal Saturn complex. You give each other structure, meaning, and purpose. You know that you both are very reliable, practical, hardworking partners that organize your activities to efficiently carry out your long term plans for

6
Moon-Sign (Rashi)

6.1 Predictions by Moon-Sign (Janma Rashi)

The position of Moon in a Sign at the time of birth is called the Moon-Sign or Birth Rashi or Rashi. Example: If the Moon is in the Sign of Mesh (Aries), the Moon-Sign or the Birth Rashi or Rashi (Janma Rashi) is Mesh. It is also called the Moon-Sign Chart, in which Moon in the first House. Accordingly, the distance of the house of all the planets from the Moon is accessed, which is essential to predict the effects of the Maha Dasa and Antar Dasa of planet.

Aries Rashi (Moon in Aries): Name starts with phonetic (Chu, Hey, Cho, La, Li, Lu, Ley, Lo, Ae):
The Moon in Aries is not congruent to her nature. He/She is restless, eyes, inflicted with diseases, unfaithful, gives pleasures to his wife, fears of drowning into water, hard working and is full of tranquillity in his old age. He/She is adventurous and is too changeable, moody, whimsical and flirtatious. He/She is likely to meet with all sorts of disappointments and even disillusion.

Taurus Rashi (Moon in Taurus): Name starts with phonetic (Ee, U, Aye, Oh, Va, Ve, Vo, Vay, Vo):
The Moon in Taurus is a natural domicile. He/She is charitable, pious, virtuous, wealthy, and full of radiance, good health and long lived. He/She has determination, loyal friend and is emotionally very strong and seldom changes his/her mind. He/She attracts opposite sex for strong romance. He/She takes up occupations of real estate, property, art, design, jewellery and business. He/She is an ambitious, selfish, has a goal for a luxurious home, plenty of money and intellectual but has a very practical, astute, shrewd ability to judge the average conditions in life.

Gemini Rashi (Moon in Gemini): Name starts with phonetic (Ka, Ke, Koo, Gha, Jna, Cha, Kay, Ko, Haa):
The moon in Gemini is considered weak. He/She has a melodious voice, talks sweetly, is kind hearted, very lusty and

is prone to throat diseases, famous, wealthy, fair complexioned, tall, clever, genius, of firm resolution, efficient in work, and remains judicious in every situation. He/She thrives on communication. He/She prefers job in media, travel as well as sales. He/She chooses his/her partners. He/She is highly imaginative, educated, and is both, a good teacher and a sharp student. He/She cannot limit to one activity and business at a time and would do well with a strong & practical partner. He/She will be entrusting the partner with most of the decision-making in business. He/She is very successful as news person, advertising agency, writers, authors or any other creative field. He/She is likely to make plenty of money, and enjoy great popularity. He/She is good in the business world. He/She will retain a youthful look and behaviour and succeeds at "staying young forever."

Cancer Rashi (Moon in Cancer): Name starts with phonetic (He, Hoo, Hey, Ho, Daa, Dee, Doo, Dey, Do): The Moon is considered royal in her own Cancer. He/She is wealthy, has patience, serves his teacher, very clever, lives in a foreign land, keeps good company, and has a high degree of intelligence. He/She is sympathetic, kind, compassionate and sensitive to others' feelings. He/She is with excellent memory, fond of home and parents, peaceful, gentle, affectionate, and romantic. He/She may get delayed marriage. He/She may be excellent artists, musicians, and poets and home life is very important to him/her. As a parent, he/she is quite nurturing and lavishes loved ones. He/She does not hesitate to use tears or a self-sacrificial attitude to get his/her point across. Real estate is an especially good area to invest in for him/her. He/She would also do well in a business run from home and has natural tendency to put on weight.

Leo Rashi (Moon in Leo): Name starts with phonetic (Ma, Me, Moo, May, Moo, Ta, Tee, Too, Tay): The Moon in Leo is the natural domicile of the Sun. He/She forgives easily, loves to travel, likes to eat non-vegetarian food, is full of fear, keeps good company, is humble, has excessive anger, devoted to his parents and achieves fame. He/She is self-sacrificing, generous, conservative, discriminating, encouraging, romantic, optimistic, brilliant robust, strong, and decisive and a natural leader and exudes energy and drive. He/She loves excitement and action. He/She may sacrifice everything in the cause of righteousness and justice. He/She makes others dependent on him. He/She cannot be convinced against his/her will, nor be swayed against emotions. He/She is strong-minded and determined. He/She is warm, loving and outgoing. He/She doesn't settle for less than what he/she wants and is too much commanding. He/She will be smothered with love and care, and may be henpeck.

Virgo Rashi (Moon in Virgo): Name starts with phonetic (Too, Pa, Pee, Pu, Sha, Na, Tha, Pay, Poe): The Moon in Virgo is unimpeded mind, passion and flesh. The keyword is "criticism". He/She is a sensualist, respects the virtuous people, religious, clever, charitable, poet, follower of the Vedas, lover of humanity, interested in dance and music, likes to travel, and is troubled by his/her Spouse. He/She has no confusion in the mind, strong social conscience, good communication and is sociable, logical, back-seat drivers and clear-headed. He/She has a strong personal code of conduct and set high standards. He/She frequently takes nursing, dietetics, teaching, and secretarial work as careers. He/She is considered to be cold-blooded and overly ambitious. He/She is very conscious of keeping fit, both physically and mentally. He/She also enjoys taking part in national politics or social issues. He/She makes well, stable business partners and a successful professional. He/She is grave, sexless people with no sexual curiosity, and doesn't understand the meaning of sex. He/She is exceedingly active, and will put more energy into house cleaning, attention to business, and personal doctoring than any other type of person.

Libra Rashi (Moon in Libra) (Chitra last half, Swati, Visakha first 3 quarters)

Name starts with phonetic (Ra, Ree, Ru, Ray, Tha, Thee, Thoo, They): The Moon in Libra is the best positions. The keyword is "decision". The moon in this position gives artistic temperament, creative ability, good mental understanding, but no executive ability. He/She gets angry unnecessarily, talks sweetly, has restless eyes, mixed fortunes, authority inside the house but powerless outside, a devotee and likes to travel. He/She is charming, creative, and diplomatic. He/She may be romantically amorous, notoriously fickle and wavering in romance. He/She is financially motivated, can be reckless, careless, and/or squandering. He/She is known for charm and social grace, and presents an image of total balance and harmony, and has problem of the kidneys and allergies.

He/She takes professions as law, architecture, politics, the arts, and even homemaking. He/She is attracted to very gracious partner and finds a good one in life, although it may not be the first one. He/She is voluptuous, deceptive in habits, with a voracious physical appetite. Usually he/she is so pretty and has so much charm that marriage is a foregone conclusion.

Scorpio Rashi (Moon in Scorpio): Name starts with phonetic (Tho, Na, Nee, Noo, Ney, No, Yaa, Yee, Yoo): The Moon in Scorpio is favourable position. The keyword is "Ulterior Motivation". He/She is a traveller from his childhood, has yellow eyes, lusty, proud, behaves roughly with his relatives, acquires wealth through hard work, and is wicked towards his mother. He/She is intelligent, cold, sensual, emotional, materialistic and secretive. He/She can become superb occultists and astrologers. The position is favourable for jobs in medicine, surgery, chemistry and investigative work. He/She can be very possessive, jealous, very cruel and vindictive. He/She never forgets a wrong that someone has done, and will plot and plan for years or decades, if necessary, to seek revenge. He/She is intensely emotional but projects a perfectly cool exterior at all times. His/Her frenzied desires are never satisfied within home. He/She is constantly seeking outside satisfaction. There are few women who have a good deal of scandal running through the life.

Sagittarius Rashi (Moon in Sagittarius): Name starts with phonetic (Yey, Yo, Ba, Bee, Bu, Dha, Pha, Dha, Bay): The Moon position in Sagittarius is risk-taking. The keyword is "enthusiasm". He/She is pious, wealthy and virtuous, has a loving nature, knowledge of fine arts, likes drawing and painting, has a wife full of good qualities, sweet-talker, has a heavy physique and in some rare cases, a destroyer of his family. He/She is eternal students with an urge to higher education, impulsive, blunt and outspoken, magnetic and forceful, actively philosophical and believes in justice and fair play and helps out anyone who is in need. He/She likes astrology and prophecy. He/She hates anything hidden or secret. He/She has a strong sense in gambling and believes that he/she simply cannot lose and therefore take unconsidered risks. If it is necessary to terminate a relationship, he/she does so quickly and cleanly without looking back. He/She knows something better is waiting just around the next corner. He/She is great spenders and enjoys a jolly, pagan sort of social life, uninhabited and full of romantic interest. He/She is incurable opposite sex partners' chasers and has put a great deal of enthusiasm into his/her dashing love affairs.

Capricorn Rashi (Moon in Capricorn): Name starts with phonetic (Bo, Ja, Je, Ju, Jay, Jo, gha, Ga, Gee): Moon in Capricorn is one of the least desirable positions. The key word is "Management". He/She has values in his family, is under the influence of his wife, scholar, undertakes charitable work, respects his mother, is wealthy, has obedient servants, is kind hearted, has a large family, and lot of worries also. He/She lacks sympathy, and has innate selfishness, self-preoccupation, dignity, tenacity and a realistic vision of the world. He/She has defeated ambitions and dreams, misfortunes, occupational and financial troubles, credit difficulties and all sorts of other misfortunes. He/She will have positions of executive, administrative, public and organizational positions and commercial pursuits. He/She has a natural desire to rise to a position of power and fame and is willing to work hard for accomplishments. He/She is not fortunate enough to achieve fame and fortune and becomes terribly frustrated and may even develop ill health as a result. He/She is very selfish, cautious, and thrifty and lays own aims and ambitions.

Aquarius Rashi (Moon in Aquarius): Name starts with phonetic (Goo, Gay, Sa, See, So, Say, Da): The Moon in Aquarius is fixed. The keyword is "disinterestedness". He/She is lazy, owns the most expensive vehicles, wealthy, is blessed with beautiful eyes, and has a simple nature. He acquires wealth and knowledge, achieves fame on account of his virtuosity and kindness, is fearless and enjoys his wealth. He/She is idealistic, caring for the global village, a little detached to home and paradoxically. He/She possesses integrity and honesty, and is not likely to ask for help when in trouble, but ready to help others. He/She is well liked and have strong religious and philosophical instincts coupled to a humanitarian urge. He/She possesses absence of jealousy and possessiveness, and favours all forms of humanitarian, political, and educational pursuits, exploration in all fields, authorship, and astrology too. He/She has a tendency to gossip and spread rumours. He/She loves the business and professional world, and has plenty of patience, kindness, intelligence and understanding.

Pisces Rashi (Moon in Pisces): Name starts with phonetic (Dee, Du, Tam, De, Do, Cha, Che): The Moon in Pisces is not favourable position. The key word is "anxiety". He/She is brave, talks cleverly, but often has excessive anger, loved by his family, a devotee, a very fast walker, efficient in charity and knowledge, virtuous, sacrificing and receives affection and love from his friends and family members. He/She has strong creativity, powerful imagination and impressionability. He/She is natural worshippers of beauty, very loyal to friends and is more inclined to romantic attitudes. He/She succeeds best in intuitive judgement, discretion, assiduity, and detailed work. He/She does well as entertainers, dealers with liquids of all sorts, promoters, seafarers, and detectives. He/She has strong intuitive and psychic qualities. He/She should generally follow his intuition, and cares a great deal about others and seek to serve society as a whole in own way. He/She gives love freely to others, and may get deceived by others. He/She may feel like rejecting the world altogether. His/Her life probably will not be one of the Cinderella or Prince Charming of his/her dreams. He/She is not afraid to work hard and is subjected to wild swings in mood. He/She can give pure, unselfish, transforming love and compassion. He/She can often find satisfaction and relief in religion and art.

7

Synastry (Vedic)

7.1 Synastry by Nakshatra Kuta

We will deal with the question of Partners Synastry and the importance of the various Kutas or physiological and psychological junctions in the human body and how the consideration of each Kuta would enable us to appreciate the harmony or discord likely to prevail between intending Partners.

(i) Mahendra Kuta: Mahendra Kuta is a kind of Varan Kuta matching without points. This refers to happiness together. It gives blessings, well-being and longevity in the Partners life. If the one partner is born in the 4th Nakshatra from that of the other, it is Mahendra; if he is born in the 7th, it is known as Upendra. Mahendra gives wealth and Upendra gives prosperity. The constellation of the first partner is counted from that of the second partner and should be the 4th, 7th, 10th, 13th, 16th, 19th, 22nd or 25th. This combination promotes well being and longevity of the partners.

(ii) Rajju Kuta: This indicates the strength or duration or partnership and therefore it merits special attention. Nakshatra are classified into 6 categories. If the Nakshatra of the partners do not fall in the same category, it is auspicious. If the Nakshatra fall in the same category, the results indicated as per, Kantha – Loss; Kati- Poverty; Pada- Distance, Siro- end; and Kukshi – Loss in partnership. The Janma Nakshatra of the partners should not fall in the same Rajju. If they fall in Sira (head) end is likely; if in Kantha (neck) the unhappiness; if in Udara (stomach) unhappiness; if in Kati (waist) poverty may ensue; and if in Pada (foot) the one may be always wandering. Hence, it is desirable that the partners have constellations belonging to different rajjus or groups for safety in life. In other places Rajju is referred to as 'Rope Agreement'. The 27 stars are to be arranged in three avenues, going three steps forward and three steps backward continuously as shown in the chart given below:

Star Number	Star Number	Star Number
1	2	3
6	5	4
7	8	9
12	11	10
13	14	15
18	17	16
19	20	21
24	23	22
25	26	27

Application of this chart to the stars of the one and other is considered very important. If the two stars be found in the same avenue, it would be the blemish called Sama-Rajju. So the alliance should not be recommended. If they are found in the central avenue, it would be called Madhyama-Rajju, and extremely harmful

Body Part	Nakshatra1	Nakshatra 2	Nakshatra 3
Kantha (Throat)	Rohini	Aridra	Hasta
Kati (Waist)	Bharani	Pushya	Purva Phalguni
Pada (Feet)	Aswini	Aslesha	Magha
Siro (Head)	Mrigashira	Chitra	Dhanishta
Kukshi (Navel)	Krittika	Punarvasu	Uttara Phalguni

Body Part	Nakshatra4	Nakshatra 5	Nakshatra 6
Kantha (Throat)	Swati	Sravana	Satabhisha
Kati (Waist)	Anuradha	Purva Ashadha	Uttara Bhadrapada
Pada (Feet)	Jyeshta	Moola	Revati
Siro (Head)	--	--	--
Kukshi (Navel)	Vishakha	Uttara Phalguni	Purva Bhadrapada

(iii) Vedha Kuta: This means affliction. It provides obstacles in partnership. Even if there is other agreement such, Vedha will prevail. The Nakshatra of the partners should not fall in the same group of Vedha. Certain constellations are capable of affecting or afflicting certain other constellations situated at particular distances from them. For instance, Aswini is said to cause Vedha to 18th constellation (viz. Jyeshta) from it; Bharani to the 16th (viz Anuradha) and so on. The following pairs or constellations affect each other and therefore no partnership should be brought about between the partners who's Janma Nakshatra belong to the same pair unless there are other relieving factors.

That pair are Aswini and Jyeshta; Bharani and Anuradha; Krittika and Visakha; Rohini and Swati; Aridra and Sravana; Punarvasu and Uttarashadha; Pushyami and Poorvashadha; Aslesha and Mula; Makha (Magha) and Revati; Purva and Uttarabhadra; Uttara and Purvabhadra; Hasta and Satabhisha, Mrigasira and Dhanishta. In our example the constellations of the couple (Mrigasira and Dhanishta) belong to the prohibited pair. Vedha is found in these pairs, and so Vedha Kuta is absent.

"Literally this word (Vedha) means piecing or hurting. A particular star is said to be Vedha or pierced by another. Where there is such a Vedha-relationship between two stars, alliance should not be recommended. There are five kinds of Vedha as (1) Kantha-Vedha, (2) Kati-(Uru)-Vedha, (3) Pada-Vedha, (4) Shiro-Vedha and (5) Kukshi-Vedha. They are so called because they hurt or pierce the neck, hips or thighs, feet head and stomach respectively. The 27 stars (Nakshatra) are grouped under the five heads as follows:

(1)   Kantha-Vedha: Rohini, Ardra, Hasta, Svati, Sravana, Shatabhishakam.

(2)   Kati-(Uru)-Vedha: Bharani, Pusya, Purvaphalguni, Anuradha, Purvashadha, Uttarashadha, Uttarabhadrapada.

(3)   Pada-Vedha: Aswini, Ashlesha, Magha, Jyestha, Mula, Revati.

(4)   Shiro-Vedha: Mrgashirsha,   Chitra,   Dhanistha.

(5)   Kukshi-Vedha: Krttika, Punarvasu, Uttaraphalguni, Vishakha,   Uttarashadha,   Purvabhadrapada.

If the stars of the partners occur in the first group, there would be Kantha-Vedha (neck piercing), Vedha leading to death of either partner. If it be in the second group, it would be Kati-Vedha (piercing of the hips or thighs) whose effect is poverty. If in the 3rd group, Pada-Vedha is resulting in wandering from place to place, poverty or loss of position. In the 4th group it is Shiro-Vedha resulting in the business end. In the last group, Kukshi-Vedha this would bring about destruction of business. Hence the stars of the partners should not belong to the same group. In the absence of this blemish alliance would be beneficial. This is one way of finding Vedha.

(iv) Gotra Kuta: It provides prosperity of the lineage after partnership. Gotra Agreement: "Ancient Sages who handed down the Vedic lore, are seven in number; Marici, Atri, Vasishtha, Angiras, Pulastya, Pulaha and Kratu. There are many lists of ancient sages who are called progenitors of Gotra like Vishvamitra, Kashyapa, Gauttama, Jamadagni and Bharadvaj. This lineage (gotra) of horary sages is based on two factors; birth and learning, i.e. from father to son; and teacher (mentor - preceptor) to disciple.

If the partners belong to the same Gotra Nakshatra, it brings disaster to the lineage. If the Gotra is different, then the lineage shall prosper. It is said that identical Gotra of partners would lead to calamity. If they belong to different Gotra, their union will lead to happiness, prosperity and prosperity. This subject can be looked into from another angle. If the stars obtained from the longitudes of their respective Lagna belong to the same Gotra or different ones, their effects would be in order moderate or excellent.

Gotra	Nakshatra 1	Nakshatra 2
Marichi	Aswini	Pushya
Vashishtha	Bharani	Ashlesha
Angirasa	Krittika	Magha
Atri	Rohini	Purva Phalguni
Pulastya	Mrigashira	Uttar Phalguni
Pulaha	Aridra	Hasta
Kretu	Punarvasu	Chitra

Gotra	Nakshatra 3	Nakshatra 4
Marichi	Swati	Abhijit
Vashishtha	Vishakha	Sravana
Angirasa	Anuradha	Dhanishta

Atri	Jyeshta	Satabhisha
Pulastya	Moola	Purva Bhadrapada
Pulaha	Purva Ashadha	Uttar Bhadrapada
Kretu	Uttara Ashadha	Revati

(v) Vihanga Kuta: It provides dominance of one partner over other. If the Nakshatra of the partners indicate the same bird, it is auspicious. Otherwise, one bird shall dominate over the other in the descending order of Peacock, Cock, Crow, Pingala, and Bharandhaka.

Birds	Nakshatra 1	Nakshatra 2	Nakshatra 3
Bharandhaka	Aswini	Bharani	Krittika
Pingala	Aridra	Punarvasu	Pushya
Crow	Uttar Phalguni	Hasta	Chitra
Cock	Jyeshtha	Moola	Purva Ashadha
Peacock	Dhanishta	Satabhisha	Purva Bhadrapada

Birds	Nakshatra 4	Nakshatra 5	Nakshatra 6
Bharandhaka	Rohini	Mrigashira	--
Pingala	Ashlesha	Magha	Purva Phalguni
Crow	Swati	Vishakha	Anuradha
Cock	Uttar Ashadha	Sravana	--
Peacock	Uttar Bhadrapada	Revati	--

(vi) Bhuta (Nakshatra Element) Kuta: It provides Psychological nature in the personality. If the Nakshatra of the partners are of the same Bhuta, it is favourable. If they are different, the results are as follows: Fire + Air are Favourable. Earth + Other are Favourable. Water + Fire are bad.

Element (Bhuta)	Nakshatra 1	Nakshatra 2	Nakshatra 3
Earth	Aswini	Bharani	Krittika
Water	Aridra	Punarvasu	Pushya
Fire	Uttara Phalguni	Hasta	Chitra

| Vayu | Anuradha | Jyeshtha | Moola |
| Akasha | Dhanishta | Satabhisha | Purva Bhadrapada |

Element (Bhuta)	Nakshatra 4	Nakshatra 5	Nakshatra 6
Earth	Rohini	Mrigashira	--
Water	Ashlesha	Magha	Purva Phalguni
Fire	Swati	Vishakha	--
Vayu	Purva Ashadha	Uttara Ashadha	Sravana
Akasha	Uttara Bhadrapada	Revati	--

(vii) Bhuta (Rashi Element) Kuta: It provides Spiritual nature in the personality.

Element (Bhuta)	Lord	Rashi 1	Rashi 2	Rashi 3
Earth	Mercury	Gemini	Virgo	--
Water	Venus/Moon	Taurus	Cancer	Libra
Fire	Mars/Sun	Aries	Leo	Scorpio
Vayu	Saturn	Capricorn	Aquarius	--
Akasha	Jupiter	Sagittarius	Pisces	--

(viii) Rasi Kuta: It provides Mental Compatibility. Principle 1: if the Janma Rasi of one partner falls in the sign which, in the Moon's Ashtakvarga of the other partner chart, has more Bindus, it is good. Similarly, if the Janma Rasi of one partner falls in a sign which, in the Moon's Ashtakvarga of the other, has more Bindus, it is equally good. Principle 2: Find out the Kaksha in which the Moon is found in the one horoscope. If the Janma Rasi of the one falls in the sign of the lord of this Kaksha, the compatibility is good. The reverse is also holds true. (Kakshya: Each sign is divided into 8 Kaksha of 3d 45m each lorded by Saturn, Jupiter, Mars, Sun, Venus, Mercury, Moon and Lagna). If the Rasi of one happens to be the 2nd house from that of the other and if the Rasi of the other happens to be the 12th house from that of the one, all kinds of evil, unwanted results will follow. But if the Rasi of the partner falls within the 12th from the other or the Rasi is the 2nd from that of the one it foresees longevity for the partnership. If the Rasi of the one is the 3rd from the other there will be misery and sorrow. But if the Rasi of the one is the 3rd from the other there will be happiness. If the one's Rasi falls in the 4th from that of the other, then there will always be great poverty; but if the Rasi of one happens to be in the 4th from the other there will be great wealth. If the one's Rasi falls in the 5th house from the other unhappiness can be expected. And if the one's Rasi is in the 5th house from the others, there will be enjoyment and prosperity. Where the Rasi of the partners are both in the 7th houses mutually, then there will be health agreements and happiness. If the one's Rasi falls in the 6th house from the others there will be a loss of prosperity.

The number of units for Rasi Kuta is 7 (seven).

Exception: When both the Rasi are owned by one planet or if the lords of the two Rasi happen to be friends, then any malefic effects attributed to the inauspicious placement of any planets is cancelled out.

(ix) Upapada Kutta: It provides the images of the partners related to each other and hence a strong say on the partner's hip affairs and harmony. Upapada is the Arudha of the 12th house in any horoscope. This is computed by counting as many houses from the 12th lord as the lord gained from the 12th house. If the Upapada falls in the 12th or the 6th house, then count 10th from the sign again to arrive at the final Upapada. Upapada tells a lot about the partner, who is committed to follow you for the whole lifetime. This shows how the images of the partners are related to each other and hence has a strong say on the partnership affairs and harmony. The matching criteria for Upapada is given here, such as, 1. The Lagna of one should be in trine or 7th from the Upapada or in the Paka Upapada and vice versa. 2. The Upapada and AL should be placed in Kendra or Trine or 3/11 to each other. Otherwise, this shows lack of harmony in the relationship. 3. The 2nd of Upapada rules the longevity of the partnership. If the 2nd house or the lord is afflicted by malefic such as Rahu, Ketu, then the partnership can be in serious troubles. This is also true if the lord is debilitated and aspect by malefic. 4. The remedy of all partnership troubles is to fast on the day ruled by

7.2 Synastry by Varan Kuta (Nakshatra Caste):

It provides mental, ideals and principles between the partnerships. It indicates the degree of Spiritual mentality of two partners, harmony in thinking and ideas, and a balance between the types of mentality between two partners to minimize ego problems so that they will be well tuned to each other and will execute all work with harmony and affection. This seems to signify the degree of advancement of spiritual development, and application of the partners. If Rasi is Pisces, Scorpio and Cancer it represents the highest development - Brahmin; Leo, Sagittarius and Libra indicate the second grade - or Kshatriya; Aries, Gemini and Aquarius suggest the third grade or the Vaishya; while Taurus, Virgo and Capricorn indicate the last grade, viz., Sudra. One belonging to a higher grade of spiritual development should not enter into partnership with others of lesser development. The vice-versa or both belonging to the same grade or degree is allowed. It would be excellent, if the stars of both be of the same class. Castes of Signs are given below:

(1) Brahmins: Cancer, Scorpio, and Pisces.
(2) Kshatriyas: Aries, Leo, and Sagittarius.
(3) Vaishyas: Taurus, Virgo and Capricorn.
(4) Shudras: Gemini, Libra and Aquarius.

Synastry of horoscopes in respect of Jati is considered on the basis of the lunar signs of the partners. Rules that have been stated for the castes of lunar mansions (stars) should be applied to the Rashi

7.3 Synastry by Vaishya Kutta (Janma Rashi Synastry):

It provides dominance power, affection and love between the couple. The Vaishya indicates the degree of Peace and Harmony, the intensity of love and mutual attraction, a power struggle in the partnership or the power equation. This indicates the degree of magnetic control or amenability the partner would be able to exercise on the other. It is based on the sign occupied by the Moon (Rashi). Principle: If the Janma Rashi of the one is the Vasya Rashi of the other Janma Rashi or vice versa; or if the one's Rashi happens to be the Vasya Rashi of the others, the one shall dearly love the others and vice versa. The affection among them is the strongest. The Vasya Rasi for different Rasi is given hereunder.

This is important as suggesting the degree or magnetic control or amenability of the partner would be able to exercise on the other. "Should the male's natal sign (lunar) be Vyasa (docile) to the female's lunar sign, the alliance would lead to a happy conjugal life. In the list given above some Rashi are mutually Vashya, not all; e.g. Gemini are mutually docile, while Leo and Libra are not mutually so. Though Sagittarius and Pisces are owned by one and the same planet, yet there is no mutual Vayshyatva (docility) between the two signs. In judging compatibility of the two charts, you have to start also from the one's Rashi, Nakshatra etc. From that point of view if the one's Rashi be Vashya to the others sign, it should be considered as a point of agreement.

Example: In the following cases, the Rashi of partners becomes the Vasya Rashi of each other, such as, Gemini – Virgo – This is 4th – 10th from each other; and Cancer – Scorpio – This is 5th – 9th from each other.

Rashi (Moon Sign)	Vasya Rashi (Moon Sign)
Aries	Leo; Scorpio
Taurus	Cancer; Libra
Gemini	Virgo
Cancer	Scorpio; Sagittarius
Leo	Libra
Virgo	Pisces; Gemini
Libra	Capricorn; Virgo
Scorpio	Cancer
Sagittarius	Pisces
Capricorn	Aries; Aquarius
Aquarius	Aries
Pisces	Capricorn

7.4 Synastry by Dina Kuta (Tara) (Nakshatra Tara Synastry):

Tara Kuta indicates fortune and misfortune in the partnership. It checks the Longevity of partnership so that partnership remains healthy and live a lifetime of each other, to ensure no financial or emotional instability in partnership due to the loss of a partner or ill health and well-being of the partners after partnership.

Method 1: Dina Kuta means mating of charts or compatibility for 'day to day living and sharing of happiness and sorrow'. The word Dina simply means day and refers to the day to day living and sharing. Dina Kuta is based on the Nava-Tara chakra of the female chart. In the Nava-Tara chakra, the 27 Nakshatra are divided into three groups of nine each, starting from the Janma Nakshatra. The first group of 9 is called as Janmarsha, the 10th to 18th Nakshatra from the Janma Nakshatra is known as the Karmarsha group and the 19th to 27th Nakshatra is known as the Adhanarksha. The transit of the natal Moon on the various Nakshatra is judged to give favourable and unfavourable results to the mind and its perception of the various events. One's Nakshatra in the 3rd (Vipat), 5th (Pratyak) or 7th (Naidhana) from the Janma Nakshatra of the other are considered very unfavourable and is harbinger of troubles and shows separation. The 6th (Saadhana) Nakshatra is also inauspicious, however, not as bad as the 3rd/ 5th or 7th. This is the basis for the Dina Kuta matching. The point is that when two partnerships, they should be supportive of each other during the days of worry and troubles. Example: On a day when the Moon is in the 3rd from Janma Nakshatra of one and he is sorrowful then this should be strong for the other and other should be able to support him to tide over the difficulties. That is the reason why Dina kuta is necessary. There are 9 categories of birth stars (Nakshatra) such as Janma, Sampat, Vipata, Kshema, Pratyari, Sadhaka, Vadha, Mitra and Ati-mitra. The counting is done from one's Birth Star to that of other and or vice versa. Count the constellations of the one from that of the other and divide the number by 9. If the remainder is 2, 4, 6, 8 or 0 it is good. If the remainder comes to 3, 5, 7, then the Tara/Dina match is considered inauspicious. Example: The constellation of one (Mrigasira in Taurus) counted from that of the other's (Dhanishta in Makar Rasi) gives 10. This is divided by 9 leaving a remainder of 1 and hence there is no compatibility between them, and the units of strength are scored zero (0) on this account.

Method 2: "The natal asterism is called Janma-Nakshatra; the 10th is termed Karma-Nakshatra and the 19th Sadhana. One whose star is the 3rd (Vipat), 5th (Pratyak) or 7th (Vadha) from the others star, is not accepted for alliance. If it be the 3rd from the 10th, Karma, star of the others, only its first quarter should be shunned, and not the other quarter. Similarly in the case of the 5th star the 4th quarter alone is bad, and other quarters are all right. In the case of Vadha (7th) star the 3rd quarter is bad, other quarters may be accepted. In case of the one's Sadhana (19th) star, only the first, fourth and third quarters respectively are to be rejected, especially when the particular quarter or Navamsha belongs to a malefic. For example, in star Aswini, the first quarter belongs to a malefic - a Papamsha - because it is owned by Mars. It is exceedingly bad to have one born in the 3rd or 7th star from the others. The 5th star however is not that bad. Should the one's birth happen in the 88th Amsa from the other's Amsa of natal star, it would be extremely harmful. Similar is one born in the 108th Amsa, i.e. the last quarter of the previous star. The good result of this agreement is long life.

Method 3: One born in the 88th or 108th Navamsa from the Navamsa Moon of the other is not auspicious. 88th Navamsa is the 4th Navamsa from the Navamsa Moon, whereas 108th is the 12th from the Moon sign. 108th Pada is just the Pada before the Moon's Nakshatra Pada; whereas the 88th Pada is 20 Padas before the other's Janma Nakshatra Pada.

Method 4: It will be better for them for their happiness, if the Janma Nakshatra of one is farther from the others Nakshatra.

7.5 Synastry by Graha Maitri (Attachment or Harmony Synastry):

It also provides Life style & objectives of life, Psychological disposition. The Graha Maitri indicates the attachment, the bond, the amount of adjustment, affection, level of affection, interactions, day-to-day behaviour, basic likes and dislikes and type of lifestyle between the partnerships. For the purposes, the planetary lords of the Moon Sign in each partner's natal chart are compared. If the lords of the Janma Rasi of both are mutual friends or are owned by one planet, the match is favourable. The mutual placement of the lords of the Janma Rasi of the partners should determine the extent of the compatibility. In all actuality this is the most important Kuta of all as it deals with the psychological dispositions of the partners, their individual mental attitudes and their affection for each other. In considering Graha Maitri the friendships between the lords of the Janma Rasis of the two people concerned is very important in determining partnership. The Planetary Relationship of Planets is given in the Table below:

Table: Planetary Relationship of Planets

Planets	Friends	Neutrals	Enemies
Sun (Surya)	Moon, Mars, Jupiter	Mercury	Saturn, Venus,
Moon (Chandra)	Sun, Mercury	Mars, Jupiter, Venus, Saturn	--
Mars (Mangal)	Sun, Moon, Jupiter	Venus, Saturn,	Mercury
Mercury (Buddha)	Sun, Venus,	Mars, Jupiter, Saturn	Moon
Jupiter (Guru)	Sun, Moon, Mars,	Saturn,	Mercury, Venus
Venus (Sukra)	Mercury; Saturn,	Mars, Jupiter	Sun, Moon
Saturn (Sani)	Mercury, Venus,	Jupiter,	Sun, Moon, Mars

According to the Shashtra it is not the temporary dispositions based on karma that the individual is going through that is to be taken into consideration, it is the birth constellation and not the birth chart as a whole.

In Vedic or devotee circles we want know primarily if the coming together of these two souls will be conducive to their spiritual development. Some who are more inclined to tasting the goods before purchase, and who are not so patient as to wait for the relationship to mature like to have trial periods of association. In all honesty it is not a substantial method to work with as we have seen, for such trials have made many errs. Previously in Vedic times, when arranged marriages were organised properly the partners were brought together each willing with an open mind and heart, knowing that to be there together this far many calculations, and great endeavour had been ensured to protect their, and their society's mutual interests. There is a science involved here, and guidelines are necessary to be followed. To circumvent proven fact, to either speed things up, or get what one likes without endeavour is not possible. Anything that is worth anything is worth working for, nothing comes cheep. When things do come cheaply, generally it doesn't last long.

When the lords of the Janma Rasi of the partners are friends, the Rasi Kuta is said to obtain full strength. When one is a friend and the other is neutral, it is passable or all right, and when both are neutral, Rasi Kuta is considered very ordinary. When both are enemies, Rasi Kuta does not exist.

Exception: Even when there is no friendship between the Janma Rasi lords of the partners, Rasi Kuta can be obtained by friendships of the planets on the Navamsa occupied by the Moon.

Example 1: In our illustration, the Janma Rasi lords are Venus and Saturn. Both are friends. Therefore the Rasi Kuta is complete. Supposing the partners is born in Makha 2 (Leo) and Satabhisha 2 (Saturn). The lords will be the Sun and Saturn respectively and they are not friends. In such a case if the Navamsa relationship is considered, then the Moon will be in Taurus (Venus) and Capricorn (Saturn) respectively. Venus and Saturn are friends and therefore the synastry is permissible. One will have to be very careful in the assessment of these factors and on superficial grounds no horoscope should be rejected as unsuitable or unfortunate.

7.6 Synastry by Gana Kuta (Temperament or Nature Synastry):

The Gana indicates the compatibility of temperaments, philanthropic psychological propensity, tolerant and religious inclination, ethics, moral and principle. It increases of affection & bond with time between the partners. The Gana is based on the Nakshatra (Constellation). This has an important bearing on the temperament and character of the perspective couple concerned. Compatibility of temperament is called for in Vedic muhurtha astrology. A difference of temperament may be harmonious and complimentary. It is said in Bhaktirasamrta Sindhu that for there to be Rasa in relationships there has to be some difference, something for one to suggest and for the other to find out. But the centrifocal point has to be at least the same so that a compatibility of temperament is there, it is essential for the satisfaction of any partnership including a peaceful union. Three Gana (temperaments of nature) are to be taken into consideration - Deva or divine, Manusha (human), and Rakshasa which is termed in Shashtra as diabolical, difficult, or even demoniac. Deva represents piety, goodness of character and charitable nature. Manusha is a mixture of good and bad, while the Rakshasa suggests dominance, selfishness and sometimes violence. These different natures are indicated by their birth constellation (Janma Nakshatra). Distaste for piety and religion etc., cannot be easily compatible with that of a religious and pious person it is obvious. A difference in beliefs and values cannot be overbalanced or set right by sexual compatibility. Thus accordingly, one whose karma is that he is born in this world with the nature of Rakshasa Gana probably will not get on well with the person of Deva Gana.

"According to astrological calculation, a person is classified according to whether he belongs to the godly or demoniac quality. In that way the spouse was selected. One of godly quality should be attached to other of godly quality. Then they will be happy. But if one is demoniac and other godly, then the combination is incompatible; they cannot be happy in such a marriage. A Deva can match a Deva, a Manusha can match a Manusha, and a Rakshasa can match a Rakshasa. Another point is that a Manusha or Deva should not try to match unless there are neutralising factors. If match is performed between prohibited Gana there will be constant quarrels and disharmony, the partners would then welcome an opportunity for extra disputes or separation.

The following Nakshatra are Deva Gana: - Punarvasu, Pushyami, Swati, Hasta, Sravana, Revati, Anuradha, Mrigasira, and Aswini. Manusha Gana is Rohini, Purva, Purvashadha, Purvabhadrapada, Bharani, Ardra, Uttara Phalguni, Uttarashadha, and Uttarabahdrapada. Rakshasa Gana is Krittika, Aslesha, Magha, Chitra, Vishakha, Jyeshta, Mula, Dhanishta, and Satabhisha.

7.7 Strength of Partnership Bond

The Ascendant and the Planets' position in it indicate the strength of the bond between the Partners such as, Unsatisfactory (US) or Satisfactory (S) or very good (VG) with respect to bond as given in the Table.

Legend: M - indicates 1st parner and F - indicates 2nd partner. S - Indicates Satisfactory and US – indicates Unsatisfactory VG – indicates Very Good. Taur – indicate Taurus; Gemi – indicate Gemini; Canc – indicate Cancer; Scorp – indicate

Scorpio; Sagit – indicate Sagittarius; Capri – indicate Capricorn; Aquar – indicate Aquarius.

Table 10: Strength of Bond with Ascendant and Planets in twelve Signs

Sign	Ascendant		Sun		Moon		Mars		Mercury	
	M	F	M	F	M	F	M	F		
Aries	US	US	US	US	--	--	--	S		
Taurus	S	S	VG	VG	-	--	--	US		
Gemini			--	--	--	--	--	--		
Cancer		US	--	--	VG	US	--	--		
Leo	US	US	--	--	US	VG	US	--		
Virgo			US	US	--	--	US	--		
Libra			US	US	--	--	US	US		
Scorp.	US	US	--	--	--	--		S		
Sagit.			--	--	--	--	S	--		
Capri.	S	S	--	--	--	--	S	--		
Aquar.	S	S	--	--	US	US	S	S		
Pisces	S	S	--	--	--	--	S	S		

Table 11: Strength of Bond with Ascendant and Planets in twelve Signs

Sign	Jupiter		Venus		Saturn		Rahu	
	M	F	M	F	M	F	M	F
Aries	--	--	--	--	VG	VG	--	--
Taur	VG	VG	VG	VG	--	--	VG	VG
Gemi	--	--		US	VG	VG	--	--
Canc	US	US	--	--	--	--	--	--
Leo	--	--	US	US	--	--	--	--
Virgo	--	--	US		--	--	--	US
Libra	US	US	US	US	US	--	--	US
Scorp.					--	US	--	--

Sagit.	US	VG	VG	VG	--	US	--	--
Capri.	--	--	--	--	--	US	VG	VG
Aquar.	--	--	VG	VG	--	--	US	--
Pisces	US	US	VG	VG	--	--	VG	VG

Table 12: Strength of Bond with occupation of Planets in twelve Houses

Bhava	Sun		Moon		Mars		Jupiter	
	M	F	M	F	M	F	M	F
1ST	US	--	VG	VG	US	US	VG	US
2ND	--	US	US	US	VG	--	VG	US
3RD	--	--	--	--	VG	--	--	--
4TH	VG	VG	--	--	--	--	--	--
5TH	US	US	--	--	US	--	--	US
6TH	--	--	--	--		VG	--	US
7TH	VG	VG	--	--	US	US	US	VG
8TH	VG	VG	VG	--	--	--	US	US
9TH	VG	VG	--	US	VG	--	--	--
10TH	--	US	US	VG	VG	VG	--	--
11TH	VG	VG	--	--	--	--	--	--
12TH	--	US	VG	VG	--	--	US	US

Table 13: Strength of Bond with occupation of Planets in twelve Houses

Bhava	Venus		Saturn		Rahu	
	M	F	M	F	M	F
1ST	--	US	US	US	--	US
2ND	--	US	--	--	VG	VG
3RD	VG	VG	--	VG	--	--
4TH	--	--	VG	--	--	--
5TH	US	US	--	--	S	VG
6TH	--	VG	--	--	VG	VG
7TH	--	--	--	--	US	US
8TH	VG	VG	--	--	VG	US
9TH	VG	VG	--	--	VG	VG
10TH	US	US	VG	--	US	US
11TH	--	--	--	--	US	US
12TH	VG	VG	US	US	--	US

8

Sun-Sign

8.0 Synastry by Sun-Sign

The date of birth determines the Sun-Sign and gives special attributes as mentioned below:

Sun Sign: Sun Sign is the active part of your personality and shows itself with blaring intensity, behaviours, and the personality traits. Moon Signs convey our shadow selves; personality traits, but shown through our Sun Signs are bold and clear.

Table: Sun Sign with Date of Birth

Sun Sign	From	To	Element
Aries	March 21st	April 19th	Fire
Taurus	April 20th	May 20th	Earth
Gemini	May 21st	June 20th	Air
Cancer	June 21st	July 22nd	Water
Leo	July 23rd	August 22nd	Fire
Virgo	August 23rd	September 22nd	Earth
Libra	September 23rd	October 22nd	Air
Scorpio	October 23rd	November 21st	Water
Sagittarius	November 22nd	December 21st	Fire
Capricorn	December 22nd	January 19th	Earth
Aquarius	January 20th	February 18th	Air
Pisces	February 19th	March 20th	Water

8.1 Synastry by Aries (March 21st - April 19th):

Who works from morning to evening, and never likes to be outdone?

Who is outspoken, alert, and ambitious and whose walk is almost like a run?

Personal Quality: The Arian is vital, impulsive, born leader, adventurous, brave, fearless, highly dominating, and full of energy, aggressive, argumentative, good athletes and soldiers and makes enthusiastic lovers. He/She is outspoken, alert and quick to act and speak. He/She is always willing to help the persons in need. He/She is not a follower. He/She is large hearted and speak straight forward for what he/she feel about the person. He/She is childishly egocentric, extremely demanding and liable to throw tantrums if denied. He/She is quick to anger and known for his impatience, and is prone to be arrogant. Under planetary afflictions he is subject to brain fever, dizziness, nosebleed, neuralgia, inflammation of the cerebral hemispheres, and diseases of face.

Positive Quality: He/She is generous, a lover of justice, and wishes to earn by own efforts and never looks at others wealth. He/She gives time, effort, money and sympathies to others. He/She likes to be challenged and enjoys solving any obstacle. He/She has both moral and physical courage.

Negative Quality: He/She is not very tactful in communicating and will never bend, has strong feeling of admirations and is not diplomatic and is sharp tongue and shows anxiety. He/She has a spending nature to maintain the image. He/She gets nervous when things are not moving his/her way. He/She is quick-tempered, violent, impatient, egotistical and intolerant.

Physical Appearance: He/She is angular, slim in early life, although may fill out later. He/She has sharp elbows and knees.

Relationships: He/She is possessive, jealous, faithful and idealistic, passionate and incurably romantic. He/She tends towards joyous sex and close relationships throughout the lives. His/Her life partner will be Leo and Libra. Aquarius and Sagittarius might be very helpful to him/her in business.

Career: His/Her income and social status will rise at the age of 48-52 and will get promotion at the age of 30, 36 and 45. He/She will deal with 'futures' on the money market or hacking through the Amazonian forest. He/She can be Dentist, Director, E M T, Entertainer, Entrepreneur, Landlord, Lawyer, and Make-up artist, Optometry, Producer, Sports person and Stockbroker.

Health: He/She is always in a tearing hurry and often has a fast metabolism, which keeps the weight down. He/She is prone to stress and suffer from tension headaches.

Lucky stone: Coral & Pearl. The glittering Coral will gives all the courage, makes him/her rich and gives a comfortable future. Topaz and Moonstones will be auspicious too.

Lucky Number: The number 1, 9 & 14 can bring luck in life.
Lucky Colour: Revel in the magic of peacock blue or Shades of Red will be lucky.
Lucky Day: Dig gold on Tuesday.
Lucky Flower: Sweet Pea.

8.2 Synastry by Taurus (Vrishabha) (April 20th - May 20th):

Who loves good things and smiles through life except when crossed?

Who is stolid, tenacious and determined and thinks he knows the most?

Personal Quality: The Taurus is stable, stubborn, generous, highly reliable, practical, ambitious, and good in the position of managers and achieves almost everything in life. He/She makes friendships very rarely but, once made, he/she is faithful. He/She is reliable, responsible, affectionate and loyal. He/She is easy to get along with and good team player. He/She can reach to the desired height with hard work, devotion and patience. He/She is practical, reserved and is possessing tremendous willpower and self-discipline. He/She is incredibly and uncompromisingly loyal. His/Her economic position will be good from the age of 35-46 and becomes a rich man in the society. His/Her early part of life is very struggling. His/Her children are generally intelligent and bright and he/she has a pleasant married life.

Positive Quality: He/She is helpful and does a lot of things for family and considers it as a sacred duty. He/She has ability to concentrate and never leaves anything unfinished and does not believe in shortcut of anything. He/She is warm, loving, gentle and charming most of the time. He/She is honest, reliable and loving.

Negative Quality: He/She very seldom changes the mind once made-up and does not bother at all for the result. He/She is expressed in dullness, stubbornness and resistance to change. He/She is very suspicious and is afraid of getting deceived. He/She believes to the person so easily that any body can cheat him/her. He/She cannot forget or forgive people so easily.

Physical Appearance: His/Her stature varies from short to medium to stocky. His/Her eyes are bright and soulful and he/she carries himself gracefully.

Relationships: He/She is intense and passionate. He/She demands perfection from mate and is exacting. This makes him/her ardent and fascinating lovers. He/She makes charming company and is loyal and devoted.

Career: He/She has a wide spread of potential careers, right from banking to the fine arts. He/She will hold on to one job for the rest of his/her working lives. Here are some occupations that a he/she might consider such as Advertising director, Antique dealer, Business person, Cashier, Clothing designer, Financial advisor, Florist, Patron of the arts, Perfumer, Real estate agent, Singer, Venture capitalist and Woodworker.

Health: Traditionally, he/she is endowed with a vigorous constitution and splendid health. If there is a weak point, it is usually his/her throat or neck. He/She is hopelessly addicted to food and alcohol.

Ideal Partner: He/She vibes best with one of Taurus.

Lucky stone: Diamond, Emerald & Blue Sapphire. Wearing Diamond or Emerald or Blue Sapphire can bring wealth and makes a better person and can give him/her strength.

Lucky Number: Number2, 3 & 8 are best for good fortune.

Lucky Colour: Lotus pink or Shades of verdant green will do a world of good.

Lucky Day: Monday is the day of new beginnings

Flower: Daisy

8.3 Synastry by Gemini (Mithuna) (May 21st - June 20th):

Who oscillates, communicates and changes often and who is fond of life, fun and pleasure? Who has a free soul and loves others attentions and exchanging of an intellectual nature?

Personal Quality: The Gemini is adaptable, dual natured, affectionate, courteous, kind, generous, scientists, and talented, bright, witty, entertaining army personnel, thoughtful towards poor and sufferer, adaptable and adjusting nature. He/She has unpredictable temperament. He/She is very well with Aquarius. He/She has wealth in later half of life. He/She will completely change his/her mind like Chameleon.

Positive Quality: He/She is usually quite, creative and has a strong self-confidence. He/She is often the centre of attraction in the gathering and is versatile, adaptable, and inquisitive and always moves along with the times.

Negative Quality: He/She is sharp-tongued and sometimes boring. He/She cannot concentrate well at one point and hence does not finish the job. He/She becomes cynical, biting, moody and quickly angered. He/She is superficial, restless, nervous, lacks concentration and conniving. He/She does not keep their promises.

Physical Appearance: He/She has small, narrow hands and feet and slim. He/She is generally tall. He/She is highly energetic and exudes oodles of charisma.

Relationships: He/She is emotionally undemonstrative but enjoys being in a lively family, and seeks a partner with strong opinions. He/She is keen to make relationships but tends to be too egocentric.

Career: He/She is particularly suited to media work, in sales pitches and can be writer, Broker, Commentator, Concierge, Correspondent, Debater, Impersonator, Journalist, Librarian, Linguist, News commentator, Novelist, Orator, and Playwright.

Health: He/She has the ill effects of smoking since he/she has delicate lungs.

Ideal Partner: He/She is best with Aquarians. Virgo, Libra and Aquarius people may help him/her.

Lucky Gem: Blue Sapphire, Diamond & Emerald. He/She can count on the Emerald, which will bless him with all the intelligence he needs.

Lucky Number: Number 3, 7 & 9 will bring him good news.

Lucky Colour: Sky Blue; Reach for the skies with sky blue.

Lucky Day: Thursday; all his dreams come true on Thursday.

Lucky Flower: Rose

8.4 Synastry by Cancer (Karka) (21st June – 22nd July):

Who cannot stick to any adhesion and changes like a season? Who most perplexing character and let's go without any reason?

Personal Quality: The Cancer is protective, jealous, sensitive, full of suspicion, not easy to understand for his/her moods, often fluctuate from sweet to cranky and lacks faith in others. He/She can be untidy, sulky, devious, and inclined to self-pity because of an inferiority complex but always ready to cooperate. He/She is very fond of food and is usually hearty eaters. He/She gets ancestral and sudden properties after the age of 35. He/She is the most family-centred persons and fiercely protective of loved ones. He/She possesses strong paternal and maternal instincts.

Positive Quality: He/She is loving, kind, faithful, loyal, honest, hard working and sensitive and leaves his unforgettable impression on others.

Negative Quality: He/She is sulky, devious, moody, clinging, manipulative, overly emotional and insecure and inclined to self-pity and prone to a sense of personal inferiority and believes his/her views, opinions and behaviour to be impeccable, and beyond question or criticism.

Physical Appearance: He/She is small with round faces and possesses a tendency to gain weight. He/She has abundant shiny hair, expressive eyes and is economical with his/her gestures.

Relationships: He/She treasures emotional bonds and doesn't severe tie easily. He/She clings on to a failing relationship and finds it difficult to let go. He/She wears his/her hearts on his/her sleeve, and is prone to emotional excesses.

Career: He/She is best suited for counselling and charity work, good journalists, writers or politicians, archaeologist, caterer, dairy farmer, deep-sea diver, dietician, hotel worker, manufacturer, merchant small businesses. He/She works well with people and often adopts the role of a mediator.

Health: He/She has a weak digestive system and constantly suffers from heartburn and ulcers. He/She tends to become hypochondriacs.

Ideal Partner: He/She seeks steady, stable and practical partners, and usually bonds best with Capricorn.

Lucky Gem: Yellow sapphire, Pearl & Coral. If he suffers with sleepless nights, a pearl could be the perfect cure. Wearing the pearl ensures peace of mind, and brings all the good luck in the world.

Lucky Number: Number 4 & 6 are his/her pick for good fortune.

Lucky Colour: white, Soak in the elegance of white.

Lucky Day: Wednesday; it is time to look ahead on Wednesday.

Lucy Flower: Larkspur.

8.5 Synastry by Leo (Simha) (July 23rd to August 22nd):

Who praises all his kindred and expects others to praise them too?

Who possess grace, dignity and generosity but cannot see their senseless view?

Personal Quality: He/She is creative, strong willed, self-confident, generous, warm hearted, loving, broad-minded and faithful. He/She has a powerful presence of mind and power of success in the conquests. He/She is called the true kings of the zodiac. He/She is manager because he dislikes subordination. He/She can be good as chairman or director in business because he is excellent organizers. He/She has self-confidence, alertness and hard struggling power. He/She never forgets the goal and tries hard to achieve it with patience, wisdom and hard labour. He/She is a dominant, always busy with some planning and work and cannot seat idle even for an hour. He/She is able to attain the top most position. He/She achieves full success after the middle age. He/She prefers position and honour to money. For position, he/she can forget any money. He/She always cares for others and other's interest or benefits except, where it is necessary to take care for him/he. He/She is born to lead. He/She is sometimes great saints. Many great Saints have born under the Leo sign. There is a reality in his/her love and he/she can do anything to please that he/she likes most. He/She will keep improving day by day after 30 years of age. He/She has a small family and their children are very brilliant and intelligent. He/She is straightforward and uncomplicated individuals. He/She is stubborn, and may suffer from short bouts of depression. He/She walks forward always, head held proudly and face turned towards the sun. He/She accumulates good amount of money, wealth and properties. His/Her fortune is good and he/she will never lack money in life.

Positive Quality: He/She is a good leader, generous and kind hearted, and confident, ambitious, loving, and honest and creative minded. He/She possesses a strong positive nature and doesn't shrink from any adverse circumstances. He/She can never bear dishonesty and injustice in life. He/She is witty and sets examples for others to follow. He/She is direct and to the point.

Negative Quality: He/She sometimes thinks himself/herself to be the only capable person in the total world. He/She believes in commanding only and does not care for others feelings. Some Leos think for money and profit and forget their other duties. It is very difficult to face his/her furies. He/She can be too sensitive to personal criticism, and when his dominance is threatened he can go into a sudden rage. He/She is conceited and arrogant.

Physical Appearance: He/She has a distinguished stature and attracts attention easily. He/She normally possesses healthy skin and a well-sculpted body.

Relationships: He/She makes wonderful social companions and is passionate and faithful lovers, but very sensitive and gets hurt easily.

Health: He/She suffers from back and spinal problems, has tendency to be overweight by lack of exercise. He/She gets easily stressed and can suffer from heart ailments.

Ideal Partner: Hence, he/she gets best with Aquarius, but Taurus, Gemini and, Sagittarius women are ideal partner for him/her.

Lucky Gem: Ruby and Topaz. The ruby is a miracle stone, and wearing it will bring him/her health and good fortune. He/She will also acquire the power to make instant decisions. The Ruby and Topaz can give him/her success and power in life.

Lucky Number: Number 1, 3, 4, 5, 6 & 9 will pull him/her out of trouble.

Lucky Colour: Saffron. Let saffron lift his/he spirits.

Lucky Day: Friday. Preen, for he/she will rake in accolades on Friday.

Lucky stone: Peridot.

Flower: Gladiolus

8.6 Synastry by Virgo (Kanya) (August 23rd to September 22nd):

Who criticizes all she sees and would even analyse a sneeze? Who is observant, shrewd but hugs and loves her own disease?

Personal Quality: The *Virgo is* critical, precise, easygoing, reliable, steady, helpful, intellectual, studious, logical, methodical, communicative, sciences, languages, takes a romance to new heights, good followers and the best employee one can ever have. He/She often dislikes delegating. He/She knows how to please the persons in power and position to get his/her work done easily and so, gets promotion very fast. He/She has a pleasant nature, colourful personality and sharp mind with a great sense of responsibility. He/She is **precise, refined, and a lover of cleanliness, hygiene and order.** It is not so easy to measure the depth of his nature. He/She knows to mould others in his shape by his clever activity. If somebody offends him/her, he/she does not show his real feelings on his/her face but act secretly to take revenge and hit back all his/her might when he/she gets the opportunity. The early part of his/her life is spent in struggles. The luck favours him/her at the age of 24, 36 to 42 years of age. He/She gets properties after the age of 40.

Positive Quality: He/She is good planners, practical and hard working, trustworthy and able to do perfect work. He/She is a hard worker, conscientious and perfectionists. He/She is plain spoken and is able to express well. He/She leads a moderate life and does not like excesses.

Negative Quality: He/She is sometimes very critical and thinks himself/herself all in all. He/She has a penchant for turning molehills into mountains, difficulties into stress and cleanliness into obsessive behaviour and a capacity for endless worry. He/She is miser and selfish and wish others to follow them.

Physical appearance: He/She has good bone structure and is often highly photogenic. He/She is attractive, with beautiful eyes that sparkle with intelligence.

Relationships: He/She is truthful and loyal. He/She is tense in close relationships, which could badly upset his/her sex lives and makes it hard for him/her to become truly intimate with those he/she loves.

Career: He/She is usually happy working in a job that calls for precision, a shrewd mind, and logic. Dogged, analytical and intellectual, he/she is makes skilled and inspired research scientists, analysts or even literary critics. He/She can be at the best as a manager, a secretary, a lawyer and a trader.

Health: He/She is often martyrs to stomachs, and may suffer from Irritable Bowel Syndrome, from food allergies. Virgo rules the abdominal region, intestines, and the lower lobes of the liver, the spleen, the duodenum, and the sympathetic nervous system.

Ideal Partner: He/She gets attracted to opposites and vibe well with dreamy Pisces.

Lucky Gem: Pearl, Topaz and Emerald. He/She should wear Emerald to crack a tough problem, to help him. The stone will bless him/her with all the intelligence he/she needs.

Lucky Number: 2, 3, 5, 6, 7 & 15are for all the good luck.

Lucky Colour: Get close to Nature. Wear earthy browns.

Lucky Day: His/Her quest for perfection pays off on Friday.

Lucky Flower: Lavender

8.7 Synastry by Libra (September 23rd to October 22nd):

Who is easygoing, sociable and keeps you waiting for half the day?

Who puts you off with promise gay and compromises all the way?

Personal Quality: The Libra is diplomat, impartial, sociable, cheerful, charming and sensitive to the needs of others. He/She is affectionate, polite in behaviour, cooperative, peace loving and sacrificing. He/She is natural arbitrators and diplomats, justice, honest and hard worker. He/She can win over his/her staunchest enemies with the help of his/her sweet voice. He/She has an idealistic and generally peace loving nature. He/She is the most civilized of the twelve signs. He/She has financial stability in the life. He/She is sure to have properties of his/her own but keep away from others' properties. He/She is more interested in making friends than enemies, and is willing to go along with others and do whatever it takes to maintain a relationship. He/She is a sensual lover and does not like any interference in the matters of love and marriage. Discord makes him/her totally insecure, and uncomfortable. He/She likes harmony in his/her life, and will do whatever it takes to have it.

Positive Quality: He/She always maintains a sweet relation with others. He/She is usually sympathetic, kind, loving nature and artistic. He/She does not like injustice, quarrels and disagreements. He/She is fair minded and loyal and have reach taste or good sense of humour. He/She does not hurt other person's feelings and likes to help the person in need.

Negative Quality: He/She is insincere and jealous and likes self admirations. He/She does not have argument power well even he is right. He/She tries to keep away from truth and painful experience. He/She can be frivolous, flirty and quite shallow. He/She is fickle minded, dependent, indecisive, sulking, and likes peace at any cost.

Physical Appearance: He/She is smart and attractive. He/She has sweet open faces with laughing eyes, and a devastating smile. He/She tends to be plump rather than angular or skinny.

Relationships: He/She highly understands his/her companions and he meets her with his own innate optimism. His/Her married life is delightful with the Gemini and Cancer girls. He/She gets special co-operation from Gemini, Aquarius, Sagittarius and Leo natives. He/She is not very good at handling relationships.

Career: He/She gets success in life as a businessman, engineer, lawyer, chartered accountant or a doctor. He/She is attracted to careers in the luxury trades, including fashion, beauty and design. He/She also makes good diplomats and counsellors. He/She is successful as writers, composers, fashion designers, interior decorators, critics, administrators, lawyers, and in civil services.

Health: This sign rules the kidneys, the lumbar region of the spine, the skin, the urethras, which are the tiny ducts running between the kidneys and the bladder, and the verso-motor system. He/She tends to suffer from nervous stomachs and ulcers. He/She needs to drink plenty of water in order to flush out the toxins from kidneys.

Ideal Partner: He/She gets along with the best ones and the gentle that captivate attention for life.

Lucky Gem: Diamond and Blue Sapphire. He/She does love the real thing, and wearing a diamond can bring him/her wealth and can make him/her a better person.

Lucky Number: Number 1, 2, 4, 7 & 21 will fill with joy.

Lucky Colour: Royal Blue. He/She is can rule over the world with royal blue.

Lucky Day: Tuesday. Pack your bags for a holiday on Tuesday.

Lucky Flower: Aster.

8.8 Synastry by Scorpio (Vrishchika) (October 23rd to November 21st:

Who has an intense and powerful nature and keeps ready an arrow in his bow? Who is self centred, wilful, proud, and detective and if you prod

him,   lets   it   go?
Who is a fervent friend, a subtle foe?

Personal Quality: Scorpion is ruthless, mysterious, magnetically, attractive and emotional intimacy and is the most intense and passionate. He/She also has immense degree of willpower and is highly tenacious. He/She is of a secretive, timid, retiring nature, one that does not talk of his affairs. He/She is honest and independent nature person with hard struggles in life. He/She does not believe in accumulating the illegal wealth. He/She has strong will power and a natural quality of leadership. He/She does not believe in empty promises. He/She can be vindictive, dangerous enemies and possesses a strong streak of venom. He/She starts good earning at the age of 24 and after 40 years of age acquire properties. He/She is self contained and self centred and seethes and doesn't give up the enmity. He/She may burst any moment. He/She is too demanding, too unforgiving of faults in others. He/She is very jealous. He/She is the symbol of sex and passionate lovers.

Positive Quality: He/She is brave, courageous, sincere and loving, subtle, imaginative, powerful, generous, loyal, passionate, exciting, and magnetic. He/She is the person with the fixed mind and achieve goal directly by his deed. He/She is not afraid of obstacles because he has a strong will power.

Negative Personality: He/She is proudly, over sensitive and careless. He/She is jealous, resentful, obstinate, compulsive, obsessive, brooding, secretive, revengeful, possessive, and extremist and can appear cold and impassive.

Health Concerns: He/She is prone to ailments of the liver and kidneys, stones and gravel in the bladder or genitals, and other genital ills such as pianism, abscesses, boils, carbuncles, fistulas, piles, ruptures and ulcers.

Physical Appearance: He/She is always striking and has a magnetic face and dress elegantly. There is a strange mysticism and magnetism in his personality, which is enchanting to the beholder.

Relationships: He/She likes to stay away from his/her family and leads an independent life. His/Her marital life with Pisces, Taurus, and Virgo and Cancer girls will be pleasant.

Career: He/She is traditionally associated with jobs such as mining and detective work. He/She makes demanding bosses. He/She might consider job such as Analyst, Criminologist,

Detective, Doctor, Enforcer, Hypnotist, Insurance agent, Investigator, Lab technician, Private investigator, Psychiatrist, Psychologist, Researcher or Scientist.

Ideal Partner: Scorpios fits best with Taurus. This sign gives him the material and emotional security he craves.

Lucky Gem: Coral, Opal. He gathers courage from the coral. The stone could also make him/her rich and also gets confident.

Lucky Number: Discover the magical powers of Number 2, 3, 7, 8 & 9.

Lucky Colour: Midnight Blue, Maroon. Kiss those blues away with midnight blue.

Lucky Day: Sunday. There will be sudden windfall on Sunday.

Lucy Flower: Chrysanthemum.

8.9 Synastry by Sagittarius (Dhanu) (November 22nd to December 21st):

Who loves the dim, religious light and always keeps a star in sight?

Who is versatile enterprising and an optimist, both gay and bright?

Personal Quality: T*he Sagittarius is m*oralistic*,* impulsive, full of versatility and eagerness, and has a positive outlook towards life. He/She enjoys travelling and exploration. He/She is ambitious and optimistic, honourable, honest, trustworthy, truthful, generous and sincere with a passion for justice and truth. He/She has very charming voice and benevolent personality. He/She is 'Yes' man and never says 'No' to anyone. He/She doesn't get demoralized in his failure. He/She does not like to harm any person and does social work too. He/She is noted for longevity, intuitiveness and original thinkers. He/She cannot gain or be successful in Gambling, Races and Stock-exchange businesses. He/She can be successful in business of white and artistic gift items or textiles and metal. He/She earns wealth after 36 year of age and also gets parental property.

Positive Quality: He/She is honest, tolerant, and friendly and trusts and respects people. He/She is kind and forgives the people easily is never proud.

Negative Quality: He/She is indiscipline and never learns even from his/her mistake in the past. He/She likes gambling and loose money. He/She does not keep his promises and does not have foresightedness and hence, he/she is unsuccessful. He/She has a quick temper and a biting tongue. Physical Appearance: He/She has darting and piercing eyes that are always likely to flash with laughter. While not particularly fashion-conscious, but he/she looks trendy.

Relationships: He/She has a happy family life and Gemini or Arian girls will be suitable as a partner and also for success in his life. Leo and Libra can be helpful for him. He/She is tactless and can hurt with his/her brutal remarks.

Career: He/She is successful in social administration, in public relations, as scientists and as musicians inquisitive. He/She works best with a tactful, organized business partner. Here are some occupations that he might consider, such as, Academic, Adventure travel guide, Advisor, Astronaut, Consultant, Entrepreneur, Inventor, Humanitarian work, Market researcher, Senator.

Health: He/She often fails to look before where he leaps, and as a result suffer quite a few bruises, pulled muscles and broken bones.

Ideal Partner: He/She needs to spend his life with organized and tolerant people, so his vibe best with Aquarius or Libra.

Lucky Gem: Topaz. If he/she is feeling invincible, thank the Hessonite for it. This is a stone of power, and the world is for you to conquer.

Lucky Number: He should be sure of success with Number 2, 3, 5, 6 & 8.

Lucky Colour: Red. Wear Reds for warmth and energy.

Lucky Day: Thursday. An old friend will brighten up an otherwise dreary Thursday.

Flower: Holly

8.10 Synastry by Capricorn (Makar) (December 22nd - January 19th):

Who takes what's due and climbs and schemes for wealth and place?

Who is confident, **a resourceful** and morns his brothers fall from grace?

Personal Quality: The Capricorn is miserly, very ambitious, self-confident, loyal, snobbish, true workaholic and accepts hard work and reaches to greater heights. He/She makes superb administrators, and often raises to very high positions in his/her careers. He/She wants to get everything with money power. He/She can be temperamental and moody. He/She acquires wealth, dominant position and what else due to his/her smooth talking and hard working. **He/She** never gives up the thing what he/she had decided to get and takes rest only after completion of the work. He/She knows the value of money only and is always ready to face the consequences to grab the money. He/She saves money and does not spend without need. He/She has very wide circle of friends. He/She is resourceful, practical manager, works well in a disciplined environment. He/She can be frugal, possessing the ability to achieve results with minimum effort and expense. He/She manages several projects simultaneously. His/Her bright carrier begins after the age of 24 and between the ages of 35 to 48 years. He/She is among the wealthy persons. He/She moulds himself/herself as per the situation and hence gets success in life.

Positive Quality: He/She is honest, very practical person and can be relied on. He/She does duty sincerely and finishes the work with great responsibility. He/She is goal oriented. He/She has great control and authority. He/She is loyal to intimates. Never impetuous, he/she considers business and personal relationships carefully before becoming involved. He/She is family person, and family usually comes first, except where business is his/her primary concern.

Negative Quality: He/She is sometimes too bossy and very narrow-minded and thinks highly of him. He/She doesn't function well in subordinate positions. He/She can spread gloom and tension in a minute and is quite capable of depressing everyone else around him. Never really up, but often down, he/she needs a positive environment to enliven his/her spirits. He/She thinks as the wisest man in the world and likes to show the others. Sometimes, he/she is very proudly. He/She is stubborn, overbearing, unforgiving, inhibited, fatalistic, condescending.

Physical Appearance: He/She tends to be tall and have sharp, angular features. He/She generally has serene expressions and an air of tranquillity. He/She takes great care of appearance.

Relationships: He/She can be surprisingly passionate behind closed doors. However, he/she takes a while to warm up and is very cautious about relating to others. He/She is very unhappy with emotional scenes and upheaval, gets hurt easily, but thaws just as quickly if he/she finds that his/her partner is genuinely repentant. Cancer and Libra girls will be ideal for his/her life partnership. Taurus, Virgo and Libra people will be helpful to him.

Career: He/She is usually determined and ambitious, with a strong sense of discipline and a good head for business. He/She sees duty and law enforcement as of paramount importance. Here are some occupations that he/she might consider such as Administrator, C E O, Coach, Commissioner, Economist, Governor, Industrialist, Leader, and Manager, Mountain climber, Office manager, Official, Operations manager, Organizer, President, Professional, Programmer, Proprietor or overbearing.

Health: Traditionally, Capricorn problem areas are their joints, bones and teeth. He/She needs to ingest extra calcium, cod-liver oil and evening primrose oil supplements to help his/her flexibility.

Ideal Partner: Always seeking a loving and strong partner, he/she gets along very well with Taurus. Capricorn needs a strong, loving partner and bond best with Taurus.

Lucky Gem: Blue Sapphire, Garnet and Diamond. This blue sapphire promises him/her all the luck he/she could wish for. He/She could soon be in for a lot of money, so better brush up on his/her financial skills.

Lucky Number: Number 6, 8, 9 & 18 are his/her secret weapon for the success.

Lucky Colour: Peacock Blue. Wear peacock blue for luck.

Lucky Day: Friday. Watch things falling into place perfectly on Friday.

8.11 Synastry by Aquarius (Kumbh) (January 20th - February 18th):

Who gives to all a helping hand but bows his head to no command?

Who are inventor, genius and superman and higher laws doth understand?

Personal Quality: The Aquarian is eccentric, inventor, technical wizard, and computer and has strong convictions. He/She possesses ill health and likes to make the world a better place. He/She has spiritual bent of mind. He/She is friendly, humanitarian and original thinkers, but can be rather eccentric. He/She is **far sighted and innovative.** He/She will go out of his/her way to help when needed, but never gets involved emotionally. He/She gets respect and the dignity in the society due to his/her kind nature and service to the people. He/She is broad-minded and expects others the same. He/She is known by his own name and is not follower but makes his own path. He/She often adopts a life style that goes against the trends, because the odd and unique fascinate him/her. He/She is an active man who is always busy with some kind of mental and physical work. He/She can achieve masterly in artistry, writing, medical, management, police or intelligence work. He/She enjoys his/her own company and is recharged by this quiet time.

Positive Quality: He/She is kind hearted, honest, kind, tolerant, cool, clear, logical people. He/She always thinks for welfare of everyone. He/She is a dedicated person and never bears injustice. He/She does not like to hurt other's feelings and are very helpful to the people in need.

Negative Quality: He/She is, sometimes, not efficient planners and the work undertaken is seldom completed. Sometimes, he/she thinks himself very clever and intelligent than others. He/She is an enigma. He/She is quite aloof people and, sometimes, does not accept his/her fault.

Physical Appearance: He/She has clean-cut good looks and a ready smile that shows off his/her excellent teeth to advantage. He/She can eat any amount of junk food and still remain slim. He/She likes to wear unusual outfits.

Relationships: Physical relationship is not so important for him/her but he/she is very emotional lover and has in depth love. Gemini, Libra, Leo and Aries are the persons who will be helpful to him/her.

Career: He/She might excel in technical fields linked with electrical and radio industries. Most are hard working, even driven in his chosen field. Many choose careers like Astrology. Here are some occupations that he/she might consider such as Academic, Adventure travel guide, Advisor, Astronaut, Consultant, Entrepreneur, Inventor, Humanitarian work, Market researcher and Senator.

Health: His/Her lungs are particularly sensitive to cigarette smoke and air pollution.

Ideal Partner: The most suitable match for Aquarians is Capricorns.

Lucky Gem: Amethyst, Blue Sapphire and Hessonite. If he/she is feeling invincible, think of hessonite for him. This is a stone of power, and the world is for him/her to conquer.

Lucky Number: He/She can be sure of success with Number 2, 3, 7 & 9.

Lucky Colour: Red. Wear reds for warmth and energy.

Lucky Day: Thursday. An old friend will brighten up an otherwise dreary Thursday.

Lucky Flower: Violet

8.12   Synastry by Pisces (Meena) (February 19th to March 20th):

Who possesses a gentle, compassionate, sensitive and spiritual nature?

Who are friendly and respond to suffering, which others encounter?

Personal Quality: The Piscean is charity, anxious, self-sacrificing, gentle, patient, malleable nature and has strong intuitive powers. He/She has superb observation, concentration while listening and good grasping power. He/She has instinct for nature, beauty, travelling, luxury and pleasure. He/She is good in subordinate positions and heads of small business. He/She is the kindest and most charitable of all the signs. He/She will make many sacrifices for other people. He/She lives life in lonely. As a lover, he/she is faithful and love to dabble in the art of sexual fantasy. He/She has many generous qualities and is friendly, good-natured, kind and compassionate, sensitive to the feelings of those around them, and respond with the utmost sympathy and tact to any suffering. He/She gets ancestral properties but he/she wishes to make money by his/her own efforts. Horseback riding, dancing, skating, swimming or sailing are favoured activities.

Positive Quality: He/She is versatile, intuitive and has quick understanding. He/She observes and listens well, and are receptive to new ideas and atmospheres. He/She readily adapts to change.

Negative Quality: His/Her dominant keyword is "I believe". His/Her nature tends to be too otherworldly for the practical purposes. He/She also dislikes disciple and confinement. The nine-to-five life is not for him/her.

Relationship: He/She tends to bond romantically well with Aquarians. He/She is never egotistical in personal relationships and gives more. He/She can be loving and affectionate partners for life.

Health Concerns: He/She can be threatened by anaemia, boils, ulcers and other skin diseases, especially inflammation of the eyelids, gout, inflammation, heavy periods and foot disorders and lameness. He/She is prone to all kinds of allergies and crippling headaches.

Partner: Aquarians are the best match for Pisceans and two tend to bond romantically.

Lucky Stone: Yellow Sapphire, Topaz & Coral.

Special Flowers: Water Lily, White Poppy & Jonquil

Special Colours: Pale Green & Turquoise

Lucky Numbers: 1, 2, 3, 4 & 6

Lucky Day: Friday

www.ingramcontent.com/pod-product-compliance
Lightning Source LLC
Chambersburg PA
CBHW080617190526
45169CB00009B/3216